Neural Networks and Genome Informatics

METHODS IN COMPUTATIONAL BIOLOGY AND BIOCHEMISTRY

Volume 1

Series Editor

A.K. KONOPKA

Maryland, USA

ELSEVIER

Amsterdam - Lausanne - New York - Oxford - Shannon - Singapore - Tokyo

Neural Networks and Genome Informatics

C.H. Wu
J.W. McLarty

University of Texas Health Center at Tyler
Department of Epidemiology and Biomathematics
11937 U.S. Highway 271
Tyler, TX 75708-3154
USA

2000

ELSEVIER

Amsterdam - Lausanne - New York - Oxford - Shannon - Singapore - Tokyo

ELSEVIER SCIENCE Ltd
The Boulevard, Langford Lane
Kidlington, Oxford OX5 1GB, UK

First edition 2000

Library of Congress Cataloging in Publication Data
A catalog record from the Library of Congress has been applied for.

British Library Cataloguing in Publication Data
A catalogue record from the British Library has been applied for.

ISBN: 0 08 042800 2

This book is dedicated to:

My late father, my husband, daughter, and son, for their inspirations. – CHW.

The late Allen B. Cohen, whose passion for science, generous spirit, inherent fairness and integrity and gentle leadership helped create a wonderfully productive environment in which to work. - JWM.

Preface

The resurgence of interest in artificial neural networks fortunately coincided with the emergence of new technology in molecular biology and the explosion of information about the genomes of humans and other species. Many important problems in genome informatics have been successfully addressed with artificial neural networks, and a vast literature has developed within the last two decades. The purpose of this book is to introduce molecular biologists and other informatics scientists to artificial neural network technology and terminology; to review the major neural network applications in genome informatics; to address the important issues in applying neural network technology to informatics; and to identify significant remaining problems.

Part I of this book gives an overview of applications of artificial neural network technology. Part II contains a tutorial introduction to the most commonly used neural network architectures, network training methods, and applications and limitations of the different architectures. Part III reviews the current state of the art of neural network applications to genome informatics and discusses crucial issues such as input variable selection and preprocessing. Finally, Part IV identifies some of the remaining issues and future directions for research, including integration of statistical rigor into neural network applications, hybrid systems and knowledge extraction.

Acknowledgements

This work would not have been possible without the support of the National Library of Medicine, the National Biomedical Research Foundation in Georgetown, Washington, D.C., and the University of Texas Health Center at Tyler. The authors are grateful for their helpful discussions with Dr. Hongzhan Huang at the National Biomedical Research Foundation and for the expertise and efforts of Dr. Karen Sloan and Sara Shepherd.

Cathy H. Wu
Jerry McLarty
Georgetown and Tyler, May1999

Contents

This part of the book consists of one chapter (Chapter 1) to provide an overview of the domain field, *genome informatics*, with its major research areas and technologies; a brief summary of the computational technology, *artificial neural networks*; and a summary of genome informatics applications. The latter two topics are further expanded into Part II, *Neural Network Foundations*, and Part III, *Genome Informatics Applications*.

CHAPTER 1

Neural Networks for Genome Informatics

1.1 What Is Genome Informatics?

Driven largely by the vast amounts of DNA sequence data, a new field of computational molecular biology has emerged: *genome informatics*. The study includes *functional genomics*, the interpretation of the function of DNA sequence on a genomic scale; *comparative genomics*, the comparisons among genomes to gain insight into the universality of biological mechanisms and into the details of gene structure and function; and *structural genomics*, the determination of the tertiary structure of all proteins. Thus, genome informatics is not only a new area of computer science for genome projects but also a new approach of life science.

The genome informatics research is moving rapidly, with advances being made on several fronts. Current methods are sufficiently accurate that they give practical help in several projects of biological and medical importance. Methods for gene recognition and gene structure prediction provide the key to analyzing genes and functional elements from anonymous DNA sequence and to deriving protein sequences. Sequence comparison and database searching are the pre-eminent approaches for predicting the likely biochemical function of new genes or genome fragments. Information embedded within families of homologous sequences and their structures, which are derived from molecular data from human and other organisms across a wide spectrum of evolutionary trees, provides effective means to detect distant family relationship and unravel gene functions.

As depicted in Figure 1.1, the major topics of genome informatics research include *gene recognition*, *functional analysis*, *structural determination*, and *family classification*, for the identification of genes, and the understanding of function, structure, and evolution of gene products. More specifically, the field involves identifying protein/RNA-encoding genes, recognizing functional elements on nucleotide sequences, understanding biochemical processes and gene regulations, determining protein structures from amino acid sequences and modeling RNA structures, and performing comparative analysis of genomes and gene families. In this book, we focus our discussions in three areas: gene recognition and DNA sequence analysis (1.1.1), protein structure prediction (1.1.2), and protein family classification and sequence analysis (1.1.3). Gene recognition and DNA and protein sequence analysis are the *sequence annotation* problem, which has undergone extensive reviews in the last decade, including the publication of several books

(Waterman, 1989; Gribskov & Devereux, 1991; Griffin & Griffin, 1994). Protein structure prediction is often termed as a *protein folding* problem and compared to deciphering the second half of the genetic code. The research has been reviewed in several books (Fasman, 1989; Gierasch & King, 1990; Nall & Dill, 1991; Creighton, 1992; Merz & Le Grand, 1994).

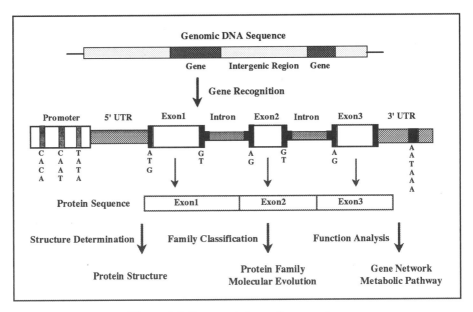

Figure 1.1 *Genome informatics overview.*
(A simplified view of gene structure is shown).

1.1.1 Gene Recognition and DNA Sequence Analysis

The problem of genomic sequence annotation involves *gene recognition* or *gene identification*, and *gene structure prediction*. It has been an active area of research in computational molecular biology for more than 15 years. The topic has been extensively reviewed (e.g., Konopka, 1994; Gelfand, 1995a; Fickett, 1996); special journal issues have been published (e.g., Bucher *et al.*, 1996); and several sets of bibliography have been compiled, such as Gelfand's FANS-REF (1995b), and Li's collections on gene recognition and DNA correlation (Li, 1999).

Gene recognition involves the identification of DNA functional elements as well as signals recognized by the transcriptional, splicing and translational machinery. The functional elements include promoters (eukaryotic and prokaryotic promoters), exons (initial, internal, and terminal exons), introns, 5' and 3' untranslated regions, and intergenic regions (Figure 1.1). It has been observed that coding DNA sequences exhibit characteristic context features, often referred to as the *content* information, that distinguish them from non-coding sequences, such as the codon usage, base composition, and periodicity (Konopka, 1994).

Signal information refers to the DNA functional *signals*, *sites* or *motifs*. They may relate to basal gene biochemistry and are common to all or most genes, such as the TATA-box and cap site in eukaryotic RNA polymerase II promoters, donor and acceptor splice sites, translation initiation and termination signals, and polyadenylation signals. In addition to these general features, there are specialized signals related to transcription or splicing. They may be complex signals for transcription initiation of specific classes of genes, and may relate to regulation of gene expression, such as transcription factor binding sites or sites recognized by other DNA- and RNA-binding proteins.

Many databases have been developed to assist the recognition of transcriptional regulatory elements. The Eukaryotic Promoter Database is a collection of experimentally characterized eukaryotic polymerase II promoters, containing sequences flanking transcription initiation sites (Périer *et al.*, 1999). TransTerm contains the coding sequence start- and stop-codon contexts, the coding sequences with associated untranslated regions, as well as several coding sequence parameters useful for detecting translational control signals (Dalphin *et al.*, 1999). Transcription Factor Database contains information pertaining to transcription factors, the proteins and biochemical entities that play a central role in the regulation of gene expression in both eukaryotes and prokaryotes (Ghosh, 1999). TRANSFAC, TRRD (Transcription Regulatory Region Database) and COMPEL are databases that store information about transcriptional regulation in eukaryotic cells. TRANSFAC/COMPEL provide views on transcription factors, their genomic binding sites, gene regulatory relationships, DNA-binding profiles, and binding site clusters (Heinemeyer *et al.*, 1999). TRRD contains experimental data on extended regulatory regions of eukaryotic genes; the regulatory elements they contain, i.e., transcription factor binding sites, promoters, enhancers, silencers, etc.; and expression patterns of the genes (Kolchanov *et al.*, 1999).

Many early approaches to the gene recognition problem focused on predicting individual functional elements in isolation, using either the *gene search by content* or *gene search by signal* method (Staden, 1984). In the search by content approach, the characteristic context features (*coding potential*) are measured by functions that calculate, for any window of sequence, a number of vectors that measures attributes correlated with protein coding function. Common examples of coding measures include the codon usage, base composition, and Fourier transform of the sequence, among which the count of in-phase hexanucleotides is found particularly effective (Fickett, 1996). Indeed, codon frequency is the most widely used feature in exon sensors (e.g., Kulp *et al.*, 1996). In the search by signal approach, functional signals are often represented by consensus sequences or position-weight matrices (Stormo, 1990a; 1990b; Waterman & Jones, 1990; Day & McMorris, 1993). For example, weight matrices have been derived for the TATA-box, CCAAT-box, GC-box, and cap signal in eukaryotic polymerase II promoters (Bucher, 1990).

More recently, a number of gene prediction programs have been developed based on the analysis and integration of multiple types of *content* and *signal* information and have taken gene structure into account. These programs include GeneID (Guigo *et al.*, 1992), GeneMark (Borodovsk & McIninch, 1993), FGENEH (Solovyev *et al.*, 1994), GRAIL (Xu *et al.*, 1994), GeneParser (Synder & Stormo, 1995), Genie (Kulp *et al.*, 1996), GENSCAN (Burge & Karlin, 1997), GeneDecoder (Asai *et al.*, 1998), and MORGAN (Salzberg *et al.*, 1998). It is well demonstrated that programs incorporating some overall model of gene structure give increased accuracy even for the recognition of individual gene components and that weighted average of all the evidence provides improved classifications.

In addition to the general content and signal information, other sequence features are considered. Burge and Karlin (1997) incorporated into their gene structure model the length distributions of initial, internal and terminal exons and introns, the dependencies between functional signals in DNA, and the various numbers of genes occurring on either or both DNA strands. Prestridge (1995) combined scores of TATA box with specific transcription elements to improve the prediction of eukaryotic polymerase II promoters. Many recent programs further utilize protein homology information derived from protein motifs or database similarity searches and show significantly improved accuracy (Snyder & Stormo, 1995; Solovyev & Salamov, 1997; Kulp *et al.*, 1997; Rogozin *et al.*, 1999). With the increasing number of completely sequenced genomes, organism-specific gene models have been developed with adjusted parameters for several model species to improve the prediction of gene structures (Solovyev & Salamov, 1997).

Gene identification typically has two phases: coding region prediction and gene parsing (i.e., determining gene structure). Separate modules (i.e., content and signal sensors) are built to capture features of the individual DNA functional elements and signals, and then combined into a gene model. The major modules are those designed for promoters, exons, introns and intergenic regions, as well as for splice sites and start sites. The coding (exon) sensors can be implemented as hidden Markov models (e.g., Henderson *et al.*, 1997), neural networks (e.g., Xu *et al.*, 1994), or decision trees (e.g., Salzberg, 1995). The splice sites and other signals can be detected using PWMs (e.g., Bucher, 1990), neural networks (e.g., Brunak *et al.*, 1991; Reese *et al.*, 1997), or hidden Markov models (Yada *et al.*, 1997). Various scoring functions have been used to combine different types of evidence and interpretation of scores, including statistical methods, such as linear discriminant function (Solovyev *et al.*, 1994), and rule-based methods (Guigo *et al.*, 1992).

Several approaches have been used for gene parsing. One common approach is to consider a set of candidate exons weighted by some statistical parameters and then construct the optimal gene, defined as a constant chain of exons using dynamic programming (Xu *et al.*, 1994; Stormo & Haussler, 1994; Snyder & Stormo, 1995; Gelfand *et al.*, 1996). However, straightforward application of this approach is conceptually difficult and computationally costly. This is overcome partially by application of neural networks either for scoring individual exons (Xu *et al.*, 1994) or in combination with a DP procedure so that multiple rounds of network training and construction of the optimal genes are performed (Snyder & Stormo, 1995).

Because nucleic acid sequences can be thought of as a language, linguistic methods are effective in understanding the gene structure (Konopka, 1994) or studying the gene expression (Rosenblueth *et al.*, 1996). Gene grammar can be used to parse DNA, and the evidence may be organized by attaching scoring rules directly to nodes of the gene parse tree (Dong & Searls, 1994). However, the components and the rules of the "DNA language" behave as though non-deterministic; hidden Markov models (Churchill, 1989) have become widely used to build stochastic models that combine the statistics and the linguistics (Asai *et al.*, 1998). Many programs have been developed using the hidden Markov model as a framework to integrate gene components into gene models (e.g., Krogh *et al.*, 1994; Kulp *et al.*, 1996).

Often multiple methods are used in the integrated system. For example, MORGAN is based on a combination of decision tree classifiers for providing scoring functions and assigning coding probabilities, Markov chain models for start and splice site recognition, and dynamic programming for optimal DNA segmentation (Salzberg *et al.*, 1998). In Genie, a generalized hidden Markov model describes the grammar of a legal parse of a DNA sequence, and probabilities are estimated for gene features by using dynamic programming to combine information from multiple content and signal sensors (Kulp *et al.*, 1997).

Gene recognition methods are often compared using the standard measures of predictive accuracy per nucleotide per exon, with the Burset and Guigo (1996) set of 570 vertebrate multi-exon gene sequences as a benchmark. For human DNA, a cross-validated standard test set of 304 genes is available (Kulp *et al.*, 1996). Still, the evaluation of integrated algorithms is complex because there is no one best interpretation of the question, "how correct is this prediction?" It is important to realize that the performance may not extend to the genomic level because the data set does not have a good representation for large, uncharacterized genomic sequences with complex structure. One should also ensure that the test sequences have low sequence similarity to training sequences.

The best program currently available can perfectly locate 95% of coding nucleotides and more than 80% of the internal coding exons, of which only 5% of the predictions do not overlap a real exon. An approximately 10% higher average accuracy of coding region recognition can be achieved by using information about similarity of predicted exons with known sequences in protein databases. Although the identification of the coding moiety of genes is satisfactory, the elucidation of gene structure and the location of promoter regions are still unreliable

1.1.2 Protein Structure Prediction

Although experimental structure determination has improved, information about the three-dimensional structure is still available for only a small fraction of known proteins. The rapid pace of genomic sequencing has further widened the gap between the number of known protein sequences and the number of experimentally determined protein structures. One of the main tasks of computational biology, thus, is to reduce this gap by predictions. The *protein folding* problem seeks to unravel the formal code that relates the amino acid sequence to its tertiary structure. The information for the three-dimensional structure must be contained within the amino acid sequence, for, under physiological conditions, the native conformation is adopted spontaneously. Thus, the classic folding problem can be simply stated: *given an amino acid sequence, predict its three-dimensional structure*. It has, however, proven to be an elusive goal. It is a difficult problem to resolve due to the size and complexity of proteins, the large number of residues that comprise a typical protein, and the structural diversity of the 20 amino acids that can occur at each residue. With an average of m equally probable conformations per amino acid residue and n residue in the protein, the total number of possible conformations will be m^n. It is estimated that up to 8^{100} different conformations are possible for a random coil polypeptide consisting of 100 amino acid residues.

Instead of asking the protein folding question of "what structure is specified by a particular sequence?" researchers have asked an alternative *inverse folding* question of "what sequences are compatible with a structure?" The inverse folding has also been variously termed as *fold recognition, threading, 3-dimensional profiles*, and *sequence-structure compatibility searches*. The fold recognition approach assumes that many different sequences fold in similar ways and there is a relatively high probability that a new sequence possesses a previously observed fold. Various approaches to fold recognition have been reviewed and the recognition process discussed (Fischer *et al.*, 1996a). The main components of fold recognition are the library of target folds, the representation of the folds, the scoring function used to evaluate the compatibility between the probe sequence and a target fold, the algorithm used to search for the optimal alignment of the sequence to each fold, and the ranking of compatibility scores for sequence-fold pairs and the assessment of the ranking's significance. To assess the performance of fold recognition methods, a benchmark, consisting of a set of protein sequences matched by superposition to known structures, was developed (Fischer *et al.*, 1996b).

Progress in protein secondary and tertiary structure predictions has been reviewed (Bohm, 1996; Eisenhaber *et al.*, 1995). Many attempts to predict protein structure have concentrated on predicting the elements of secondary structure because 90% of the residues in most proteins are involved in three classes of secondary structures, the α-helices, β-strands or reverse turns. *Helices* and *sheets* are termed *regular* structures because their residues have repeating main-chain torsion angles, and their backbone groups are arranged in a periodic pattern of hydrogen bonding. An α-helix is a periodic polypeptide structure with 3.6 residues per turn and a pitch of 1.5 A per residue. Sheets in proteins are almost invariably buried structures. Since the hydrogen bonds of helices can be localized to intrasegment partners, helical secondary structure is dependent mainly on local sequence information, and not on tertiary interactions. On the other hand, the formation of a β-sheet requires long-range interactions. In contrast to helix and sheet, *turns* are nonregular structures with nonrepeating backbone torsion angles. Remaining residues are often loosely classified as *random coil*.

1.1.3 Protein Family Classification and Sequence Analysis

Calculation of similarity between newly obtained sequence and archives of various types of information is perhaps the most valuable tool for obtaining biological knowledge. When similarity exists between query sequence and database entries, the method yields clues to function, as already illustrated by numerous novel discoveries. A range of algorithms is available to solve somewhat different sequence comparison problems. They range from the most sensitive but computationally intensive algorithms of dynamic programming, such as Needleman-Wunch (1970), Sellers (1974), and Smith-Waterman (1981), to the faster methods of BLAST (Altschul *et al.*, 1997) and FASTA (Pearson & Lipman, 1988). Whenever possible, the search should be carried out at the amino acid, rather than the nucleotide, level to reduce the noise of neutral mutations and improve sensitivity. These types of database searching methods are often referred to as *direct searching* or *lookup* method. This is in contrast to the *reverse searching* or *template* method. The former is based on pair-wise sequence comparisons, whereas the latter is based on comparisons of protein motifs, domains or families and often referred to as *family search* method.

Protein family classification provides an effective means for understanding gene structure and function and for the systematic studies of functional genomics. The classification approach has several desirable impacts on the annotation of newly sequenced genes. The utilization of family information improves the identification of genes, which are otherwise difficult to detect based on pair-wise alignments. The systematic family analysis assists database maintenance by making annotation errors apparent. Protein family organization provides effective means to retrieve relevant information from vast amounts of data. Finally, gene families are essential for phylogenetic analysis and comparative genomics.

Protein family relationships have been studied at three different levels, roughly categorized according to the extent of sequence conservation. *Superfamily* or *family* is used to indicate full-length end-to-end sequence similarity at the whole protein level; *domain* to indicate local similarity at the functional or folding unit level; and *motif* to indicate short local similarity at the functional or active site level. Many family databases have been compiled, including the PIR-International Protein Sequence Database with superfamily organization (Barker *et al.*, 1999); domain databases, such as Pfam (Bateman *et al.*, 1999), ProDom (Corpet *et al.*, 1999), and DOMO (Gracy & Argos, 1999); motif databases, such as PROSITE (Hofmann *et al.*, 1999), Blocks (Henikoff *et al.*, 1999), and PRINTS (Attwood *et al.*, 1999); as well as databases with integrated family classifications, such as ProClass (Wu *et al.*, 1999).

Several family search methods have been devised, including BLOCKS search (Henikoff & Henikoff, 1994), profile search (Gribskov & Veretnik, 1996), hidden Markov modeling (Eddy *et al.*, 1995), and neural networks (Wu, 1996). Correspondingly, the family information may be condensed as blocks, profiles or Dirichlet mixture densities (Sjolander *et al.*, 1996), hidden Markov models, or neural network weight matrices. Most of these family classification methods are dependent on the quality of multiple sequence alignments. A number of methods have been developed for multiple sequence alignment, such as ClustalW (Thompson *et al.*, 1994), and for phylogenetic analysis, such as Phylip. These methods are important for defining gene families as well as for studying the evolutionary relationships among genomes.

Protein family search at the motif sequence level involves the recognition of functional sites. Therefore, the same methods that apply to DNA signal recognition, such as the consensus sequence, position-weight matrix, or neural network method, are also applicable to the detection of protein motifs. In addition, *signature patterns* are often derived from aligned motif sequences to describe the conservative substitution of amino acid residues on certain motif positions, as in PROSITE database. Although motif patterns can be conveniently and quickly searched in query sequences, it may not be easy or even possible to define patterns that are both sensitive (i.e., all sites can be described by the pattern) and specific (i.e., all containing the pattern are true sites). An alternative is to generate a set of patterns for each motif with a wide range of specificities and sensitivities, as in EMOTIF (Nevill-Manning *et al.*, 1998).

1.2 What Is An Artificial Neural Network?

An artificial neural network is a parallel computational model comprised of densely interconnected adaptive processing elements called *neurons* or *units*. It is an information-processing system that has certain performance characteristics in common with the biological neural networks (Rumelhart & McClelland, 1986). It resembles the brain in that knowledge is acquired by the network through a *learning* process and that the interconnection strengths known as *synaptic weights* are used to store the knowledge (Haykin, 1994).

A neural network is characterized by its pattern of connections between the neurons (called *network architecture*) and its method of determining the weights on the connections (called *training* or *learning algorithm*). The weights are adjusted on the basis of data. In other words, neural networks learn from examples and exhibit some capability for generalization beyond the training data. This feature makes such computational models very appealing in application domains where one has little or incomplete understanding of the problem to be solved, but where training data is readily available. Neural networks normally have great potential for parallelism, since the computations of the components are largely independent of each other.

Artificial neural networks are viable computational models for a wide variety of problems. Already, useful applications have been designed, built, and commercialized for various areas in engineering, business and biology (Dayhoff, 1990; Murray, 1995). These include pattern classification, speech synthesis and recognition, adaptive interfaces between human and complex physical systems, function approximation, image compression, associative memory, clustering, forecasting and prediction, combinatorial optimization, nonlinear system modeling, and control. Although they may have been inspired by neuroscience, the majority of the networks have close relevance or counterparts to traditional statistical methods (Sarle, 1994; Bishop, 1995; Ripley, 1996), such as non-parametric pattern classifiers, clustering algorithm, nonlinear filters, and statistical regression models. Part II provides a tutorial for several neural network paradigms particularly useful for genome informatics applications.

1.3 Genome Informatics Applications

With its many features and capabilities for recognition, generalization and classification, artificial neural network technology is well suited for genome informatics studies. As the first application, the perceptron learning algorithm developed by Rosenblatt in the late 1950's was adapted to sequence pattern analysis by Stormo *et al*. (1982) in an attempt to distinguish ribosomal binding sites from non-binding sites. Most early genome informatics applications (reviewed by Hirst & Sternberg, 1992; Presnell & Cohen, 1993) involved the use of perceptron or back-propagation (Rumelhart & McClelland, 1986) networks. They were employed to predict protein secondary and tertiary structure, to distinguish protein encoding regions from non-coding sequences, to predict bacterial promoter sequences, and to classify molecular sequences. As the field continues to develop, researchers have broadened the choices of neural network architectures and learning paradigms to solve a wider range of problems. Part III discusses genome informatics applications and their design issues in detail.

1.4 References

Altschul, S. F., Madden, T. L., Schaffer, A. A., Zhang, J., Zhang, Z., Miller, W. & Lipman, D. J. (1997). Gapped BLAST and PSI-BLAST: a new generation of protein database search programs. *Nucleic Acids Res* **25**, 3389-402.

Asai, K., Itou, K., Ueno, Y. & Yada, T. (1998). Recognition of human genes by stochastic parsing. *Pac Symp Biocomput*, 228-39.

Attwood, T. K., Flower, D. R., Lewis, A. P., Mabey, J. E., Morgan, S. R., Scordis, P., Selley, J. N. & Wright, W. (1999). PRINTS prepares for the new millennium. *Nucleic Acids Res* **27**, 220-5.

Barker, W. C., Garavelli, J. S., McGarvey, P. B., Marzec, C. R., Orcutt, B. C., Srinivasarao, G. Y., Yeh, L. S., Ledley, R. S., Mewes, H. W., Pfeiffer, F., Tsugita, A. & Wu, C. (1999). The PIR-International Protein Sequence Database. *Nucleic Acids Res* **27**, 39-43.

Bateman, A., Birney, E., Durbin, R., Eddy, S. R., Finn, R. D. & Sonnhammer, E. L. (1999). Pfam 3.1: 1313 multiple alignments and profile HMMs match the majority of proteins. *Nucleic Acids Res* **27**, 260-2.

Bishop, C. (1995). *Neural Networks for Pattern Recognition.* Oxford UP, New York.

Bohm, G. (1996). New approaches in molecular structure prediction. *Biophys Chem* **59**, 1-32.

Borodovsky, M. & McIninch, J. (1993). GENMARK: parallel gene recognition for both DNA strands. *Comput Chem* **17**, 123-34.

Brunak, S., Engelbrecht, J. & Knudsen, S. (1991). Prediction of human mRNA donor and acceptor sites from the DNA sequence. *J Mol Biol* **220**, 49-65.

Bucher, P. (1990). Weight matrix descriptions of four eukaryotic RNA polymerase II promoter elements derived from 502 unrelated promoter sequences. *J Mol Biol* **212**, 563-78.

Bucher, P., Fickett, J. W. & Hatzigeorgiou, A. (1996). Computational analysis of transcriptional regulatory elements: a field in flux. *Comput Appl Biosci* **12**, 361-2. Introduction to the Special Issue on *Computational Analysis of Eukaryotic Transcriptional Regulatory*, October 1996.

Burge, C. & Karlin, S. (1997). Prediction of complete gene structures in human genomic DNA. *J Mol Biol* **268**, 78-94.

Burset, M. & Guigo, R. (1996). Evaluation of gene structure prediction programs. *Genomics* **34**, 353-67.

Churchill, G. A. (1989). Stochastic models for heterogeneous DNA sequences. *Bull Math Biol* **51**, 79-94.

Collins, F. S, Patrinos, A., Jordan, E., Chakravarti, A., Gesteland, R. & Walters, L. (1998). New goals for the U.S. Human Genome Project: 1998-2003. *Science* **282**, 682-9.

Corpet, F., Gouzy, J. & Kahn, D. (1999). Recent improvements of the ProDom database of protein domain families. *Nucleic Acids Res* **27**, 263-7.

Creighton, T. E. (1992). *Protein Folding.* W. H. Freeman and Company, New York.

Dalphin, M. E., Stockwell, P. A., Tate, W. P. & Brown, C. M. (1999). TransTerm, the translational signal database, extended to include full coding sequences and untranslated regions. *Nucleic Acids Res* **27**, 293-4.

Day, W. H. E. & McMorris, F. R. (1993). A consensus program for molecular sequences. *Comput Appl Biosci* **9**, 653-6.

Dayhoff, J. (1990). *Neural Network Architectures: An Introduction.* Van Nostrand Reinhold, New York.

Dong, S. & Searls, D. B. (1994). Gene structure prediction by linguistic methods. *Genomics* **23**, 540-51.

Eddy, S. R., Mitchison, G. & Durbin, R. (1995). Maximum Discrimination hidden Markov models of sequence consensus. *J Comp Biol* **2**, 9-23.

Eisenhaber, F., Persson, B. & Argos, P. (1995). Protein structure prediction: recognition of primary, secondary, and tertiary structural features from amino acid sequence. *Crit Rev Biochem Mol Biol* **30**, 1-94.

Fasman, G. D. (1989). Prediction of Protein Structure and the Principles of Protein Conformation. Plenum P, New York.

Fickett, J. W. (1996). The gene identification problem: an overview for developers. *Comput Chem* **20**, 103-18.

Fischer, D., Rice, D., Bowie, J. U. & Eisenberg, D. (1996a). Assigning amino acid sequences to 3-dimensional protein folds. *Faseb J* **10**, 126-36.

Fischer, D., Elofsson, A., Rice, D. & Eisenberg, D. (1996b). Assessing the performance of fold recognition methods by means of a comprehensive benchmark. *Pac Symp Biocomput*, 300-18.

Gelfand, M. S. (1995a). Prediction of function in DNA sequence analysis. *J Comput Biol* **2**, 87-115.

Gelfand, M. S. (1995b). FANS-REF: a bibliography on statistics and functional analysis of nucleotide sequences. *Comput Appl Biosci* **11**, 541.

Gelfand, M. S., Podolsky, L. I., Astakhova, T. V. & Roytberg, M. A. (1996). Recognition of genes in human DNA sequences. *J Comput Biol* **3**, 223-34.

Ghosh, D. (1999). Object oriented Transcription Factors Database (ooTFD). *Nucleic Acids Res* **27**, 315-7.

Gierasch, L. M. & King, J. (1990). *Protein Folding: Deciphering the Second Half of the Genetic Code*. American Association for the Advancement of Science, Washington, D. C.

Gracy, J. & Argos, P. (1999). Automated protein sequence database classification. I. Integration of compositional similarity search, local similarity search, and multiple sequence alignment. *Bioinformatics* **14**, 164-73.

Gribskov, M. & Devereux, J. (1991). *Sequence Analysis Primer*. Stockton P, New York.

Gribskov, M. & Veretnik, S. (1996). Identification of sequence pattern with profile analysis. *Methods Enzymol* **266**, 198-212.

Griffin, A. M. & Griffin, H. G. (1994). *Computer Analysis of Sequence Data*, vol. 24, 25. In *Methods in Molecular Biology* (ed. Walker, J. M.). Humana P, Totowa, NJ.

Guigo, R., Knudsen, S., Drake, N. & Smith, T. (1992). Prediction of gene structure. *J Mol Biol* **226**, 141-57.

Haykin, S. (1994). *Neural Networks: A Comprehensive Foundation*. Macmillan, New York.

Heinemeyer, T., Chen, X., Karas, H., Kel, A. E., Kel, O. V., Liebich, I., Meinhardt, T., Reuter, I., Schacherer, F. & Wingender, E. (1999). Expanding the TRANSFAC database towards an expert system of regulatory molecular mechanisms. *Nucleic Acids Res* **27**, 318-22.

Henderson, J., Salzberg, S. & Fasman, K. H. (1997). Finding genes in DNA with a Hidden Markov Model. *J Comput Biol* **4**, 127-41.

Henikoff, S. & Henikoff, J. G. (1994). Protein family classification based on searching a database of blocks. *Genomics* **19**, 97-107.

Henikoff, J. G., Henikoff, S. & Pietrokovski, S. (1999). New features of the Blocks Database servers. *Nucleic Acids Res* **27**, 226-8.

Hirst, J. D. & Sternberg, M. J. E. (1992). Prediction of structural and functional features of protein and nucleic acid sequences by artificial neural networks. *Biochemistry* **31**, 7211-19.

Hofmann, K., Bucher, P., Falquet, L. & Bairoch, A. (1999). The PROSITE database, its status in 1999. *Nucleic Acids Res* **27**, 215-9.

Kolchanov, N. A., Ananko, E. A., Podkolodnaya, O. A., Ignatieva, E. V., Stepanenko, I. L., Kel-Margoulis, O. V., Kel, A. E., Merkulova, T. I., Goryachkovskaya, T. N., Busygina, T. V., Kolpakov, F. A., Podkolodny, N. L., Naumochkin, A. N. & Romashchenko, A. G. (1999). Transcription Regulatory Regions Database (TRRD): its status in 1999. *Nucleic Acids Res* **27**, 303-6.

Konopka, A. K. (1994). Sequences and codes: Fundamentals of biomolecular cryptology. In: *Biocomputing: Informatics and Genome Projects*. (Ed., Smith, D.), Academic Press, San Diego, CA. pp. 119– 74.

Krogh, A., Mian, I. S. & Haussler, D. (1994). A hidden Markov model that finds genes in *E. coli* DNA. *Nucleic Acids Res* **22**, 4768-78.

Kulp, D., Haussler, D., Reese, M. G. & Eeckman, F. H. (1996). A generalized hidden Markov model for the recognition of human genes in DNA. *Ismb* **4**, 134-42. Human DNA data set: [ftp:@www-hgc.lbl.gov/pub/genesets].

Kulp, D., Haussler, D., Reese, M. G. & Eeckman, F. H. (1997). Integrating database homology in a probabilistic gene structure model. *Pac Symp Biocomput* , 232-44.

Li, W. (1999). A bibliography on Computational Gene Recognition (http://linkage.rockefeller.edu/wli/gene) and on Correlation Structure of DNA (/dna_corr).

Merz, K. M., Jr. & Le Grand, S. M. (1994). *The Protein Folding Problem and Tertiary Structure Prediction*. Birkhauser, Boston.

Murray, A. F. (1995). *Applications of Neural Networks*. Kluwer Academic Publishers, Boston.

Nall, B. T. & Dill, K. A. (1991). *Conformations and Forces in Protein Folding*. American Association for the Advancement of Science, Washington, D. C.

Needleman, S. B. & Wunsch, C. D. (1970). A general method applicable to the search for similarities in the amino acid sequences of two proteins. *J Mol Biol* **48**, 443-53.

Nevill-Manning, C. G., Wu, T. D. & Brutlag, D. L. (1998). Highly specific protein sequence motifs for genome analysis. *Proc Natl Acad Sci U S A* **95**, 5865-71.

Pearson, W. R. & Lipman, D. J. (1988). Improved tools for biological sequence comparisons. *Proc Nat Acad Sci USA* **85**, 2444-8.

Perier, R. C., Junier, T., Bonnard, C. & Bucher, P. (1999). The Eukaryotic Promoter Database (EPD): recent developments. *Nucleic Acids Res* **27**, 307-9.

Presnell, S. R. & Cohen, F. E. (1993). Artificial Neural Networks for Pattern Recognition in Biochemical Sequences. *Annu Rev Biophys Biomol Struct* **22**, 283-98.

Prestridge, D. S. (1995). Predicting Pol II promoter sequences using transcription factor binding sites. *J Mol Biol* **249**, 923-32.

Reese, M. G., Eeckman, F. H., Kulp, D. & Haussler, D. (1997). Improved splice site detection in Genie. *J Comput Biol* **4**, 311-23.

Ripley, B. D. (1996) *Pattern Recognition and Neural Networks*. Cambridge University Press, Cambridge.

Rogozin, I. B., D'Angelo, D., Milanesi, L. (1999). Protein-coding regions prediction combining similarity searches and conservative evolutionary properties of protein-coding sequences. *Gene* **226**, 129-37.

Rosenblatt, F. (1962). *Principles of Neurodynamics*. Spartan Books, Washington, D.C.

Rosenblueth, D. A., Thieffry, D., Huerta, A. M., Salgado, H. & Collado-Vides, J. (1996). Syntactic recognition of regulatory regions in Escherichia coli. *Comput Appl Biosci* **12**, 415-22.

Rost, B. & Sander, C. (1993). Prediction of protein secondary structure at better than 70% accuracy. *J Mol Biol* **232**, 584-99.

Rumelhart, D. E. & McClelland, J. L. (Eds.). (1986). *Parallel Distributed Processing*, Vols. 1 and 2. MIT P, Cambridge.

Salzberg, S. (1995). Locating protein coding regions in human DNA using a decision tree algorithm. *J Comput Biol* **2,** 473-85.

Salzberg, S., Delcher, A. L., Fasman, K. H. & Henderson, J. (1998). A decision tree system for finding genes in DNA. *J Comput Biol* **5,** 667-80.

Sarle, W. S. (1994). Neural networks and statistical models. In: *Proceedings of the Nineteenth Annual SAS Users Group International Conference*, pp. 1538-50. SAS Institute, Cary, NC. (ftp://ftp.sas.com/pub/neural/neural1.ps).

Searls, D. B. (1997). Linguistic approaches to biological sequences. *Comput Appl Biosci* **13,** 333-44.

Sellers, P. H. (1974). On the theory and computation of evolutionary distances. *SIAM J Appl Math* **26,** 787-93.

Sjolander, K., Karplus, K., Brown, M., Hughey, R., Krogh, A., Mian, I. S. & Haussler, D. (1996). Dirichlet mixtures: a method for improved detection of weak but significant protein sequence homology. *Comput Appl Biosci* **12,** 327-45.

Smith, T. F. & Waterman, M. S. (1981). Comparison of bio-sequences. *Adv Appl Math* **2,** 482-89.

Snyder, E. E. & Stormo, G. D. (1995). Identification of protein coding regions in genomic DNA. *J Mol Biol* **248,** 1-18.

Solovyev, V. V., Salamov, A. A. & Lawrence, C. B. (1994). Predicting internal exons by oligonucleotide composition and discriminant analysis of spliceable open reading frames. *Nucleic Acids Res* **22,** 5156-63.

Solovyev, V. & Salamov, A. (1997). The Gene-Finder computer tools for analysis of human and model organisms genome sequences. *Ismb* **5,** 294-302.

Staden, R. (1984). Computer methods to locate signals in nucleic acid sequences. *Nucleic Acids Res* **12,** 505-19.

Staden, R. (1990). Finding protein coding regions in genomic sequences. *Methods Enzymol* **183,** 163-80.

Stormo, G. D., Schneider, T. D. & Gold, L. (1982). Use of the perceptron algorithm to distinguish translational initiation sites in *E. coli*. *Nucleic Acids Res* **10,** 2997-3011.

Stormo, G. D. (1990a). Identifying regulatory sites from DNA sequence data. In: Structure and Methods. (Eds., Sarma, R. H. & Sarma, M. H.), Adenine Press, Guiderland, NY. pp. 103-12.

Stormo, G. D. (1990b). Consensus patterns in DNA. *Methods Enzymol* **183,** 211-21.

Stormo, G. D. & Haussler, D. (1994). Optimally parsing a sequence into different classes based on multiple types of evidence. *Ismb* **2,** 369-75.

Thompson, J.D., Higgins, D.G. & Gibson, T.J. (1994). CLUSTAL W: improving the sensitivity of progressive multiple sequence alignment through sequence weighting, position specific gap penalties and weight matrix choice. *Nucleic Acids Res* **22,** 4673-80.

Waterman, M. (1989). *Mathematical Methods for DNA Sequences*. CRC Press, Boca Raton.

Waterman, M. S. & Jones, R. (1990). Consensus methods for DNA and protein sequence alignment. *Methods Enzymol* **183,** 221-37.

Wu, C. H. (1996). Gene Classification Artificial Neural System. *Methods Enzymol* **266,** 71-88.

Wu, C. H., Shivakumar, S. & Huang, H. (1999). ProClass Protein Family Database. *Nucleic Acids Res* **27,** 272-4.

Xu, Y., Mural, R. J. & Uberbacher, E. C. (1994). Constructing gene models from accurately predicted exons: an application of dynamic programming. *Comput Appl Biosci* **10,** 613-23.

Yada, T., Sazuka, T. & Hirosawa, M. (1997). Analysis of sequence patterns surrounding the translation initiation sites on cyanobacterium genome using the hidden Markov model. *DNA Res* **4,** 1-7.

PART II

Neural Network Foundations

Artificial neural networks are versatile tools for a number of applications, including bioinformatics. However, they are not *thinking machines* nor are they *black boxes* to blindly feed data into with expectations of miraculous results. Neural networks are typically computer software implementations of algorithms, which fortunately may be represented by highly visual, often simple diagrams. Neural networks represent a powerful set of mathematical tools, usually highly nonlinear in nature, that can be used to perform a number of traditional statistical chores such as classification, pattern recognition and feature extraction.

One strength of the neural network approach over traditional approaches (the comparison to traditional methods is discussed in depth in Chapter 12) is that the diagrams used to construct the algorithms can be used in a visual, intuitive way to perform powerful procedures while sparing the user from arcane mathematical notations and from the complex computational details that are actually being performed. This advantage cannot be underestimated, especially for a practitioner whose primary training and experience are not in mathematics or computer science. On the other hand, successful application of neural networks may require considerable subject area expertise and intuition, especially in the formulation of the problem, design of the network and preparation of data for presentation to the neural network.

The purpose of the following four chapters is to build a foundation of understanding of basic neural network principles. Subsequent chapters will address issues specific to choosing the neural network design (architecture) for particular applications and the preparation of data (data encoding) for use by neural networks.

Not all neural networks are the same; their connections, elemental functions, training methods and applications may differ in significant ways. The types of elements in a network and the connections between them are referred to as the network *architecture*. Commonly used elements in artificial neural networks will be presented in Chapter 2. The multilayer perceptron, one of the most commonly used architectures, is described in Chapter 3. Other architectures, such as radial basis function networks and self organizing maps (SOM) or Kohonen architectures, will be described in Chapter 4.

Coupled closely with each network architecture is its *training* method. Training (or as it is sometimes called, *learning*) is a means of adjusting the connections between elements in a neural network in response to input data so that a given task can be performed. A trained network then contains an approximation of the information contained in the training data.   There are two fundamental kinds of learning performed by neural networks: supervised learning, and unsupervised learning.   With supervised learning, training data for which the answer is known is applied to a network.   An algorithm is provided for iteratively modifying the network so that its output eventually approximates the desired, known results.   Supervised training is not unlike teaching children arithmetic with flash cards, so that after proper training they can perform the procedures correctly on new sets of problems, without supervision.   Supervised learning is particularly useful for classification problems, such as constructing families of proteins with similar structure and function.   Supervised learning for multilayer perceptron network architectures and others will be discussed in Chapter 5.

With unsupervised learning, training data is applied to a neural network, without prior classification or knowledge of the answers.   An algorithm is provided for the network to modify itself, based on the data applied, in such a way that features in the data can be extracted automatically in an iterative manner.   An analogous situation would be to present pictures of animals to a person and ask him/her to group the animals that are similar, without supplying any information about the animals not evident from the pictures.   Unsupervised learning is particularly useful in situations where the distinguishing features of a set of data are not known in advance.   Unsupervised learning will be discussed in Chapter 5 for a particular architecture, Kohonen self-organizing maps.

CHAPTER 2

Neural Network Basics

2.1 Introduction to Neural Network Elements

2.1.1 Neurons

Neural networks consist of groups (layers) of processing units with connections between (and sometimes within) layers. The basic unit within a layer is an artificial neuron, a crude representation of a biological neuron. These units, like real neurons, have input connections (dendrites) and output connections (axons). Also like real neurons, neural network units have some form of internal processing that creates (or modulates) an output signal as a function of the input signal. However, there are two fundamental differences between real neurons and artificial neural network units. First, the output from a biological neuron is a pulse modulated signal - a series of pulses of fixed amplitude - with the frequency of pulses being changed in response to the signals coming in from the dendrites. An artificial neuron simply puts out a number (perhaps corresponding to a voltage) in response to the input signals. This number typically may take any value within a set range, usually 0 to 1. Second, the output from a biological neuron is continuously changing with time whereas its artificial counterpart changes only at discrete intervals of time, whenever its inputs are changed (see Figure 2.1).

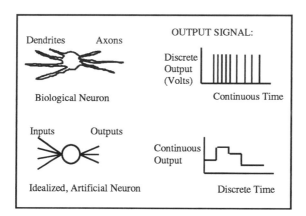

Figure 2.1 *The analogy between biological neurons and artificial neurons.*

Understanding the symbols and diagrams of artificial neural networks is essential to using neural networks. A simple neuron unit is shown in Figure 2.2 with three input signals and a single output signal. In actual usage, neurons may have dozens or even hundreds of inputs and outputs. Each neuron, shown pictorially (often as a circle, square or dot), represents a processing unit that uses all the information coming into it to produce an output value. The output value from a unit is the same for all its outputs. Connections between neurons are shown graphically by lines connecting them. A neuron unit in a neural network diagram represents a mathematical transfer function. The transfer function for some units may be a simple pass-through (e.g., multiply input by 1.0) function whereas some units represent more complicated functions such as combinations of summation, threshold functions or sigmoidal functions such as a logistic function. In mathematical terms, a neuron can be represented as

Output = F(inputs), where F is some function, to be discussed later.

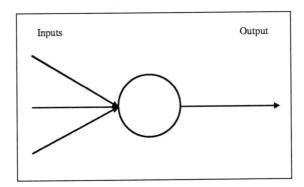

Figure 2.2 A single artificial neuron with input and output connections.

2.1.2 Connections between Elements

A set of neuronal units with connections between them constitutes a network. The connections between neurons have a signal strength, or *weight*, attached to them. A more complete picture of a simple network with input and output connections and weights is shown in Figure 2.3. Note that there are two layers to this network, with units 1 and 2 constituting layer 1 and a single unit, number 3, making up layer 2. Information flow is from left to right (in this book, information flow is assumed to be from left to right, unless otherwise specified by arrows on the connecting lines). In this simple network, the output of unit 1 is multiplied by its weight, $w_{1,3}$, as it is fed into unit 3. Similarly for unit 2, its output is multiplied by $w_{2,3}$. Unit 3 then processes this information according to its internal transfer function. By convention, the first layer, where information enters the network, is called the *input layer* and the last layer, where information exits the network, is the *output layer*. For most examples in this book, input layers are the leftmost layers and output layers the rightmost. Exceptions are usually clear from the context. Such a network is called a *feed-forward* network for obvious reasons; information flow is always in the forward direction, never backward.

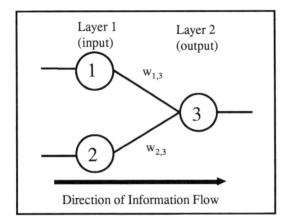

Figure 2.3 A simple feed-forward neural network.

2.2 Transfer Functions

2.3.1 Summation Operation

Practically all forms of neuron transfer functions include the summation operation, i.e., the sum of all inputs into the neuron (multiplied by their connection strength or weights) is calculated. In mathematical terms,

$$\Sigma = \sum_{i=1}^{n} x_i w_{ij} = x_1 w_1 + x_2 w_{2j} + ... + x_n w_{nj} \qquad (2.1)$$

where n is the number of input connections into unit j, and w_{ij} is the weight of the connection between unit i of the previous layer and unit j.

2.3.2 Thresholding Functions

Another function common in neurons is *thresholding*, or changing the output signal in discrete steps depending upon the value of the summation (Σ) of the input signals. The output signal of neurons can theoretically have any value between $\pm\infty$; however, typically values range between 0 and 1. Some neurons are allowed to have only the discrete values 0 or 1 (off and on, respectively), whereas others are allowed to take any real value between 0 and 1 inclusively. A simple threshold function is of the form

$$\text{Output} = 0 \text{ if } \Sigma < \theta \ ,$$
$$\text{Output} = 1 \text{ if } \Sigma \geq \theta \ ,$$

where θ is a pre-specified threshold value.

Stated another way,

$$\text{Output} = 0 \text{ if } \Sigma - \theta < 0 \ ,$$
$$\text{Output} = 1 \text{ if } \Sigma - \theta \geq 0 \ .$$

In mathematical terms, thresholding is application of the Heaviside function, $F_h(x)$ which has the value 0 when $x = \Sigma - \theta < 0$ and 1 when $x \geq 0$.

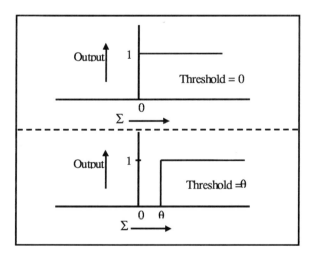

Figure 2.4 *Examples of different threshold values.*

This is illustrated for two different threshold values in Figure 2.4. Sometimes a thresholding function is referred to as a *squashing* function since a large input value is squashed into a small range of output values.

Another common name for the threshold value θ is *bias*. The idea is that each neuron may have its own built-in bias term, independent of the input. One way of handling this pictorially and computationally is to add an extra unit to the input layer that always has a value of -1. Then the weight of the connections between this unit and the neurons in the next layer is the threshold or bias values for those neurons and the summation operation includes the bias term automatically. Then the summation formula becomes

$$\sum = \sum_{i=1}^{n} x_i w_{ij} = x_1 w_{1j} + x_2 w_{2j} + \ldots + x_n w_{nj} - \theta_j \qquad (2.2)$$

or simply

$$\sum = \sum_{i=0}^{n} x_i w_{ij} = x_0 w_{0j} + x_1 w_{1j} + \ldots + x_n w_{nj} \qquad (2.3)$$

with the extra input node and bias term being subsumed in the input and weight vectors as x_0 and w_{0j}, respectively. Note that the summation operation now begins with unit 0, which is designated the bias unit and always has input value −1 (some authors use +1, but the effect is the same; the weight terms are negated).

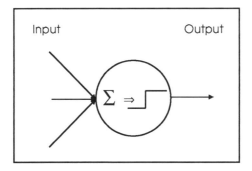

Figure 2.5 *The basic neuron model.*

So, the basic neuron can be seen as having two operations, summation and thresholding, as illustrated in Figure 2.5. Other forms of thresholding and, indeed, other transfer functions are commonly used in neural network modeling; some of these will be discussed later. For input neurons, the transfer function is typically assumed to be unity, i.e., the input signal is passed through without modification as output to the next layer; F(x) = 1.0.

2.3.3 Other Transfer Functions

The idealized step function (Heaviside function) described above is only one of many functions used in basic neural network units. One problem with this function is that it is discontinuous, and does not a continuous derivative; derivatives are at the heart of many training methods, a topic to be discussed later. An approximation to a step function that is easier to handle mathematically is the logistic function, shown in Figure 2.6 below.

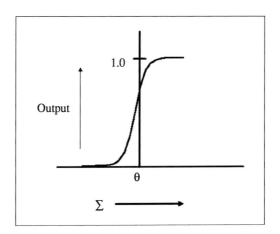

Figure 2.6 *The logistic threshold (transfer) function.*

The equation for the logistic function is

$$Output = \frac{1}{1 + e^{-\Sigma}}$$

where Σ is the weighted sum of input values as in equation 2.2 or 2.3.

Other sigmoidal functions, such as the hyperbolic tangent function, are also commonly used. Finally, Radial Basis Function neural networks, to be described later, use a symmetric function, typically a Gaussian function.

2.4 Simple Feed-Forward Network Example

To put it all together in a simple example, assume the network in Figure 2.7 has an input vector I with value

$$I = \begin{bmatrix} x_0 \\ x_1 \\ x_2 \\ x3 \end{bmatrix} = \begin{bmatrix} -1.0 \\ -0.25 \\ 0.50 \\ -1.0 \end{bmatrix}$$

In words, the bias input unit is given an input signal of value –1; the second unit, -.25; the third unit value .50; and the fourth unit, 1.0. Note that although theoretically any range of numbers may be used, for a number of practical reasons to be discussed later, the input vectors are usually scaled to have elements with absolute value between 0.0 and 1.0. Note that the bias term unit in Figure 2.7, which has a constant input value of –1.0, is symbolized as a square, rather than a circle, to distinguish it from ordinary input nodes with variable inputs.

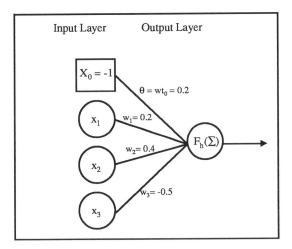

Figure 2.7 *A fully specified neural network.*

Weights can also be seen as a vector W, for this example with values

$$W = \begin{bmatrix} w_0 \\ w_1 \\ w_2 \\ w_3 \end{bmatrix} = \begin{bmatrix} 0.2 \\ 0.2 \\ 0.4 \\ -.5 \end{bmatrix}$$

Further assume that the transfer function for the input units is unity, i.e., whatever input is received is passed through without change to the next layer, $F(x) = x$. Assume a simple threshold function (Heaviside function) for the single output unit.

Computation of the output is as follows from equation 2.3 (simplified in this case since there is only one output neuron):

$$\Sigma = \sum_{i=0}^{n} x_i w_i = x_0 w_0 + x_1 w_1 + ... + x_4 w_4$$

$$= (-1)(.2) + (-.25)(.2) + (.50)(.4) + (-1.0)(-.5) = -.2 + .65 = .45$$

Since this is greater than 0, the output of the network is 1.0 , i.e., $F_h(.45) = 1$.

A different set of inputs, for example $I^T = [.5,.1,.1]$, would have yielded an output of 0 (I^T is the transpose of vector I). The same principle of information feed-forward, weighted sums and transformation applies with multiple units in each layer and, indeed, with multiple layers. Multiple layered networks will be discussed in the next chapter.

2.5 Introductory Texts

There are many excellent introductory books and journal articles on the subject of neural networks. Just a few of them are listed below in the references. Additionally, there are tutorials online at various web sites. However, the applications of neural network techniques to problems in molecular biology and genome informatics are largely to be found in scientific journals and symposium proceedings.

2.6 References

Beale, R. & Jackson, T. (1991). *Neural Computing: An Introduction*. Adam Hilger, Bristol.

Bishop, C. M. (1995). *Neural Networks For Pattern Recognition*. Clarendon P, London.

Dayhoff, J. (1990). *Neural Network Architectures: An Introduction*. Van Nostrand Reinhold, New York.

Fausett, L. (1994). *Fundamentals Of Neural Networks: Architectures, Algorithms And Applications*. Prentice Hall, Englewood Cliffs.

Haykin, S. (1994). *Neural Networks, A Comprehensive Foundation*. MacMillan College Publishing, New York.

Hinton, G. E. (1992). How neural networks learn from experience. *Sci Am* **267,** 144-51.

Wasserman, P. D. (1993). *Advanced Methods In Neural Computing*. Van Nostrand Reinhold, New York.

CHAPTER 3

Perceptrons and Multilayer Perceptrons

In the previous chapter a simple two-layer artificial neural network was illustrated. Such two-layer, feed-forward networks have an interesting history and are commonly called *perceptrons*. Similar networks with more than two layers are called *multilayer perceptrons*, often abbreviated as MLPs. In this chapter the development of perceptrons is sketched with a discussion of particular applications and limitations. Multilayer perceptron concepts are developed; applications, limitations and extensions to other kinds of networks are discussed.

3.1 Perceptrons

The field of artificial neural networks is a new and rapidly growing field and, as such, is susceptible to problems with naming conventions. In this book, a perceptron is defined as a two-layer network of simple artificial neurons of the type described in Chapter 2. The term *perceptron* is sometimes used in the literature to refer to the artificial neurons themselves. Perceptrons have been around for decades (McCulloch & Pitts, 1943) and were the basis of much theoretical and practical work, especially in the 1960s. Rosenblatt coined the term *perceptron* (Rosenblatt, 1958). Unfortunately little work was done with perceptrons for quite some time after it was realized that they could be used for only a restricted range of linearly separable problems (Minsky & Papert, 1969).

3.1.1 Applications

The simplest perceptron can be used to classify patterns into one of two classes. Training perceptrons and other networks is a numerical, iterative process that will be discussed in Chapter 5. It has been rigorously proven that training perceptrons for classification problems will converge to a solution in a finite number of steps, if a solution exists.

Hypothetical Example

To show a hypothetical example of this, consider Figure 3.1 which illustrates using two measures (say, from questionnaires) of mood to diagnose depression. The problem is to decide, solely on the basis of these two measures, whether a patient is clinically depressed or not. Note that neither measure by itself seems predictive of depression, but the two taken together show a clear dichotomy. Assume that data is available where the outcome diagnosis is known; this data is the basis for Figure 3.1 and for training the network. For a perceptron, the goal is to *learn* how to classify data into one of the two classes. In this instance, the term *learning* means that the weight terms of the network will be adjusted so that classification will be performed in some optimal manner. This is equivalent to linear discriminant analysis where the problem is to find a line (more precisely, the slope and intercept for a line) that best separates the two groups on the graph.

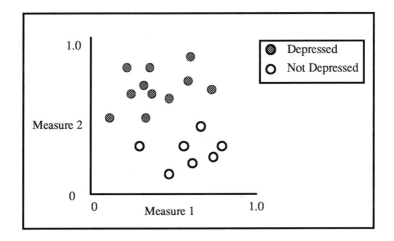

Figure 3.1 Example of a classification problem.

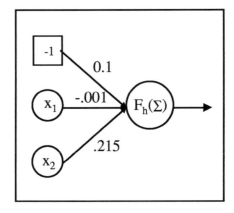

Figure 3.2. Perceptron solution of the depression classification problem.

From the figure it is easy to see that an infinite number of lines can be drawn that separate the *depressed* points from the *not depressed* points in the plane. This is a characteristic of regression and neural network applications: rarely is there one solution but a whole family of solutions for a given problem. Nevertheless, a perceptron can be easily trained to classify patients based on the two hypothetical measures posed. One perceptron with its trained weights for this particular set of data is shown in Figure 3.2.

For example if a new patient scores .5 on measure 1 ($x_1 = 0.5$) and .7 on measure 2 ($x_2 = 0.7$), then from equation 4.3,

$$\Sigma = \sum_{i=0}^{n} x_i w_{ij} = x_0 w_0 + x_1 w_1 + x_2 w_2$$

$$= (-1)(.1) + (.5)(-.001) + (.7)(.215) = 0.05$$

Since this is greater than 0, the output is 1.0, i.e., $F_h(0.05) = 1.0$; in this case the perceptron classifies the patient as depressed. The reader is encouraged to try other values and see if they agree with the diagnosis he/she would infer from the graph.

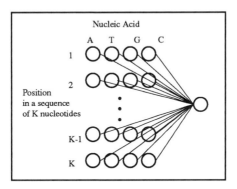

Figure 3.3 *Nucleotide matrix input to a perceptron.*

Nucleic Acid Application

For an example of a perceptron used in nucleic acid research, see Stormo *et al.* (1982). In this study, a perceptron was used to find a weighting function, which distinguishes *E. coli* translational initiation sites from all other sites in a library of nucleotides of mRNA sequences. Sequences of nucleotides were converted to a matrix; then matrices from sequences of known ribosome binding sites were presented to the perceptron along with matrices from sequences known not to be ribisome-binding sites. Using these sequences, the network was trained to classify new sequences as either binding sites or not. The application was successful in finding gene beginnings, in fact, more successful than previous approaches that were *rule-based*. This complex example has hundreds of input units, but the principles illustrated by the simple hypothetical example are the same. A matrix of values (illustrated in Figure 3.3) may be used as input, with each element in the matrix having a connection with a weight to the output unit.

Methods of encoding--converting the input data to a form usable by a neural network--will be discussed in later chapters. However, to illustrate the matrix approach shown in Figure 3.3 (as used in the cited nucleotide example) consider a nucleotide sequence of length 8: AAGATCGC. This would be represented in matrix form as

$$
X = \begin{bmatrix}
1000 \\
1000 \\
0010 \\
1000 \\
0100 \\
0001 \\
0010 \\
0001
\end{bmatrix}
$$

Each row represents a single nucleotide and the position of the 0's and 1's in a row designates the nucleotide: A = [1000], T=[0100], G=[0010], C=[0001]. It is easy to see how such a matrix could be presented to the input layer of a perceptron: each 0 or 1 bit goes into a specific input unit.

Protein Sequence Example

Schneider and others (Schneider *et al.*, 1993) applied the perceptron approach to identifying cleavage sites in protein sequences. Again, a matrix approach was used, with 12 rows representing a sequence of 12 amino acid residues (one row per residue) and four columns representing physico-chemical features of each residue: hydrophobicity, hydrophilicity, polarity and volume. The trained perceptron predicted cleavage sites correctly in 100% of test cases.

3.1.2 Limitations

Although perceptrons are quite useful for a wide variety of classification problems, their usefulness is limited to problems that are *linearly separable*: problems in which a line, plane or hyperplane can effect the desired dichotomy. As an example of a non-linearly separable problem, see Figure 3.4. This is just Figure 3.1 with an extra point added (measure 1 ≈ .8 and measure 2 ≈ .9); but this point makes it impossible to find a line that can separate the depressed from non-depressed. This is no longer a linearly separable problem, and a simple perceptron will not be able to find a solution. However, note that a simple curve can effectively separate the two groups. Multilayer perceptrons, discussed in the next section, can be used for classification, even in the presence of nonlinearities.

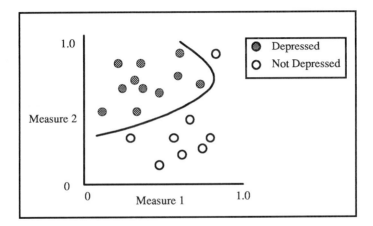

Figure 3.4 *A nonlinear classification problem.*

After the Minsky and Papert book in 1969 (Minsky & Papert, 1969) which clarified the linearity restrictions of perceptrons, little work was done with perceptrons. However, in 1986 McClelland and Rumelhart (McClelland & Rumelhart, 1986) revived the field with multilayer perceptrons and an intuitive training algorithm called back-propagation (discussed in Chapter 5).

3.2 Multilayer Perceptrons

Multilayer perceptrons (MLP) are perceptrons with more than two layers, i.e., an input layer, an output layer and at least one layer between them. The middle layers are called *hidden* layers because they are not connected directly either to network inputs or outputs. Neurons within a hidden layer are called *hidden units*. Typically all of the units in one layer are attached to all of the units of the next layer but exceptions may occur. Also, input layer units are generally not connected to output layer units, but, again, exceptions may occur. For multilayer as well as simple perceptrons, the flow of information is only in a feed-forward direction. The transfer function of units in the hidden and output layers is not the step function, i.e., Heaviside function, used for perceptron output layers, but is logistic or tanh or other sigmoid functions that have derivatives everywhere. The property of differentiability is essential to training multilayer perceptrons. The combination of nonlinear transfer functions and hidden layers allows the application of artificial neural networks to a host of nonlinear problems not solvable by perceptrons. A simple multilayer perceptron is illustrated in Figure 3.5. Note the sigmoid symbol in the hidden and output units signifying use of one of the sigmoidal transfer functions. Typically this is not shown explicitly but is assumed.

The output of a sigmoid function, such as the logistic function, is not 0 or 1 but somewhere between. Therefore a decision must be made as to what value will be called "on"; this value is called, by convention in this book, a critical value. Typically, if the output of the function is > 0.5, then the unit is said to be on (equivalent to the Heaviside function output of 1); otherwise it's off. Depending upon the application this can be changed to other critical values like 0.8 or 0.9. Higher critical values can be said to be less sensitive (it takes more input to turn them on) and lower critical values more sensitive (they turn on at lower input levels).

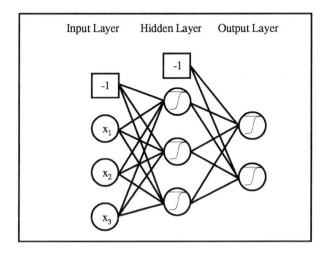

Figure 3.5 Multilayer perceptron neural network.

Also shown explicitly in Figure 3.5 are the bias units, with constant values of −1 and connections to the hidden and output units. Because of the complexity of the diagram, these are usually implied and not shown. Some very large neural networks even omit the bias terms from their neuron models (as well as from their diagrams), but for smaller units they are usually considered important. A neuron without a bias term is like the equation of a line without a constant term, which means that every line goes through the origin. Figure 3.6 shows the more common representation of the same network as in Figure 3.5, shown without input unit variable designations, without bias units and without symbols specifying the form of transfer function of hidden and output layers.

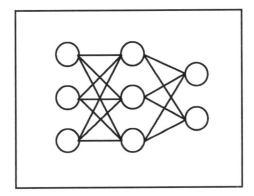

Figure 3.6 *Common representation of a three-layer perceptron.*

The three-layer network in Figure 3.6 is designated as a 3-3-2 multilayer perceptron, which specifies the number of units (excluding bias units) in each layer. Note that some authors would call this a two-layer MLP since there are two layers of <u>weights</u>. However, for this book, we will use the more common terminology, which counts layers of <u>units</u>, including the input layer. Another note concerning structure and terminology: if all the units from one layer are connected to all the units of the next layer, the network is called *fully-connected*. Fully connected networks are the norm; if a network is not fully connected, it must be made explicit which units are connected and which are not.

The number of units in the input layer is determined by the number of features used in the input vector. The number of units in the output layer is determined by the number of categories of output required. However, it is important to note that there is no best way to determine the number of hidden layers and the number of units in each layer. There is theoretical work to show that a network with sigmoid transfer functions with one hidden layer, given an appropriate number of units, can represent any desired transformation from input to output (Hornik *et al.*, 1989) within a given range of desired precision. Most network applications use a single hidden layer; however, the number of hidden units can be quite large, depending upon the difficulty of the desired task. There are heuristic methods of determining how many hidden units are needed, based on the performance of the network on training and test sets of data. For a highly theoretical discussion of these issues see the book by Ripley (1996).

3.2.1 Applications

Multilayer perceptrons have been successfully used in a multitude of diverse problems. They are particularly good at nonlinear classification problems and at function approximation.

Hypothetical Example

A multilayer perceptron with two hidden units is shown in Figure 3.7, with the actual weights and bias terms included, after training. This network can solve the nonlinear depression classification problem, presented in Figure 3.4.

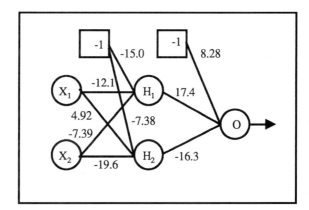

Figure 3.7 MLP solution to the example nonlinear classification problem.

To illustrate the calculations involved, consider classifying a new patient with measure 1 score of .85 and measure 2 score of .95 (X_1 =.85, X_2 =.95):

The weighted sum of all the inputs to hidden unit H_1 is

$$(-1)(-15.0) + (.85)(-12.1)+(.95)(-7.39) = -2.3055$$

The output of H_1 is calculated as

$$\frac{1}{1 + e^{-(-2.3055)}} = .0907$$

The weighted sum of all inputs to hidden unit H_2 is

$$(-1)(-7.38) + (.85)(4.92) + (.95)(-19.6) = -7.058$$

The output of H_2 is given by

$$\frac{1}{1 + e^{-(-7.058)}} = .00086$$

The weight sum of inputs to the single output unit is

$$(-1)(8.28) + (.0907)(17.4) + (.00086)(-16.3) = -6.72$$

Finally, the output of the entire network, in response to x1 = and x2 =, is

$$\frac{1}{1 + e^{-(-6.72)}} = .00121$$

Since this output value is very close to 0 (within a critical value of, say, 0.1), the output unit is considered to be "off", i.e., the patient is classified as not depressed. This fits reasonably with the nonlinear regions shown in Figure 3.4.

Examples in Genome Informatics

Tables 9.1, 10.1 and 11.1 in this book have references to many examples of applications of multilayer perceptrons to problems in genome informatics, with the particular architecture cited for each example.

A classic example is the GRAIL system for classifying DNA sequences into two groups, exons and not exons; i.e., the network is trained to identify coding regions (Uberbacher *et al.*, 1996). One version of the system uses a 13 x 3 x 1 multilayer perceptron: thirteen input units, three hidden units and one output unit. The thirteen features include sequence length, exon GC composition, Markov scores, and other complex scales describing the candidate sequence.

An example of a large-scale multilayer perceptron application with hundreds of input, hidden and output units is the protein classification system of Wu and others (Wu *et al.*, 1995). Inputs to the network included counts of amino acid pairs, counts of exchange group pairs and triplets, and other combinations (methods of encoding sequence information are discussed in detail in Chapter 7 of this book). Total input vector sizes ranged from 400 to 1,356 units. Hidden layer sizes were typically 200 units and output layer sizes varied with the number of protein superfamilies into which patterns were to be classified, from 137 to 178 superfamilies. Although this network was quite large, the principles of operation for the multilayer perceptrons are identical to those described for the simple examples given above.

3.2.2 Limitations

Multilayer perceptrons have been successfully applied to many problems in classification and function approximation. These typically are supervised learning situations where results are known in advance for the training set of data. Without training data, such that the answer or target value is known for each input vector, multilayer perceptrons are not appropriate: for example, when the task is to cluster or group objects but it is not known in advance what the clustering groups will be. Other kinds of problems for which multilayer perceptrons are not appropriate are problems in which it is necessary to learn from the entire training set and generalization is not possible (such as predicting random numbers or times series with no underlying structure). Some of the major limitations of multilayer perceptrons involve training: convergence problems and large numbers of presentations of the input data. Unlike simple perceptrons, multilayer perceptrons are not guaranteed to converge to an existing solution, and training them may be quite difficult in some situations. There are many other varieties of neural network architectures, some of which will be described below, but the most commonly used neural network is the multilayer perceptron.

3.3 References

Hornik, K., Stinchcombe, M. & White, H. (1989). Multilayer feedforward networks are universal approximators. *NEUNET* **2**, 359-66.

McClelland, J. L. & Rumelhart, D. E. (1986). *Parallel Distributed Processing*. MIT Bradford P, Cambridge.

McCulloch, W. S. & Pitts, W. H. (1943). A logical calculus of the ideas immanent in nervous activity. *Bull Math Biophys* **5,** 115-33.

Minsky, M. & Papert, S. (1969). *Perceptrons: An Introduction To Computational Geometry.* MIT P, Cambridge.

Ripley, B. D. (1996). *Pattern Recognition And Neural Networks.* Cambridge UP, Cambridge.

Rosenblatt, F. (1958). The perceptron: a probabilistic model for information storage and organization in the brain. *Psychol Rev* **65,** 383-408.

Schneider, G., Rohlk, S. & Wrede, P. (1993). Analysis of cleavage-site patterns in protein precursor sequences with a perceptron-type neural network. *Biochem Biophys Res Commun* **194,** 951-9.

Stormo, G. D., Schneider, T. D., Gold, L. & Ehrenfeucht, A. (1982). Use of the 'Perceptron' algorithm to distinguish translational initiation sites in E. coli. *Nucleic Acids Res* **10,** 2997-3011.

Uberbacher, E. C., Xu, Y. & Mural, R. J. (1996). Discovering and understanding genes in human DNA sequence using GRAIL. *Methods Enzymol* **266,** 259-81.

Wu, C. H., Berry, M., Shivakumar, S. & McLarty, J. (1995). Neural networks for full-scale protein sequence classification: sequence encoding with singular value decomposition. *Machine Learning* **21,** 177-93.

CHAPTER 4

Other Common Architectures

There are literally dozens of kinds of neural network architectures in use. A simple taxonomy divides them into two types based on learning algorithms (supervised, unsupervised) and into subtypes based upon whether they are feed-forward or feedback type networks. In this chapter, two other commonly used architectures, radial basis functions and Kohonen self-organizing architectures, will be discussed. Additionally, variants of multilayer perceptrons that have enhanced statistical properties will be presented.

4.1 Radial Basis Functions

Networks based on radial basis functions have been developed to address some of the problems encountered with training multilayer perceptrons: radial basis functions are guaranteed to converge and training is much more rapid. Both are feed-forward networks with similar-looking diagrams and their applications are similar; however, the principles of action of radial basis function networks and the way they are trained are quite different from multilayer perceptrons.

4.1.1 Introduction to Radial Basis Functions

The essence of the differences between the operation of radial basis function networks and multilayer perceptrons can be seen in Figure 4.1, which shows data from the hypothetical classification example discussed in Chapter 3. Multilayer perceptrons classify data by the use of hyperplanes that **divide** the data space into discrete areas; radial basis functions, on the other hand, **cluster** the data into a finite number of ellipsoid regions. Classification is then a matter of finding which ellipsoid is closest for a given test data point.

The hidden units of a radial basis function network are not the same as used for a multilayer perceptron, and the weights between input and hidden layer have different meanings. Transfer functions typically used include the Gaussian function, spline functions and various quadratic functions: they all are smooth functions, which taper off as distance from a center point increases. In two dimensions, the Gaussian is the well-known bell-shaped curve; in three dimensions it forms a hill.

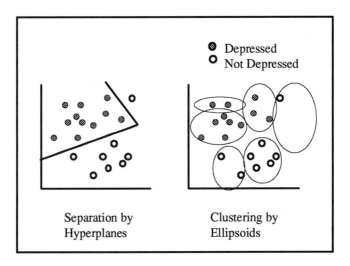

Figure 4.1 *Different approach to classification by multilayer perceptrons (left) and radial basis function networks (right).*

For radial basis function networks, each hidden unit represents the center of a cluster in the data space. Input to a hidden unit in a radial basis function is not the weighted sum of its inputs but a *distance* measure: a measure of how far the input vector is from the center of the basis function for that hidden unit. Various distance measures are used, but perhaps the most common is the well-known Euclidean distance measure.

If x and μ are vectors, the Euclidean distance between them is given by

$$D = \|x - \mu\| = \sqrt{\sum_i \left(x_i - \mu_i \right)^2} \qquad (4.1)$$

where x is an input vector and μ_j is the location vector, or center of the basis function for hidden node j. The hidden node then computes its output as a function of the distance between the input vector and its center. For the Gaussian radial basis function the hidden unit output is

$$h_j(D_j^2) = e^{-D_j^2 / 2\sigma_j^2} \qquad (4.2)$$

Where D_j is the Euclidean distance between an input vector and the location vector for hidden unit j; h_j is the output of hidden unit j and σ^2_j is a measure of the size of the cluster j (in statistical terms it is called the variance or the square of the standard deviation).

To see how this works, consider the simple radial basis function network in Figure 4.2 which has only one input unit, two hidden units and one output unit.

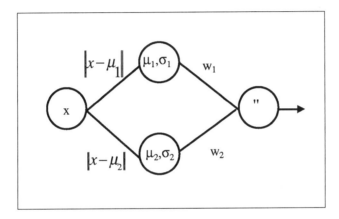

Figure 4.2 A simple radial basis function network.

In this example, μ_i and σ_i are the parameters for the hidden units. The final output of the network is a simple weighted sum of outputs from each of the hidden units and their weights. Input to the hidden units is the distance between the input x and their respective centers, μ_i . If x is closer to μ_1 than μ_2 then the output from hidden unit 1 is stronger than for hidden unit 2. Figure 4.3 illustrates this point, with a Gaussian basis function, with $\sigma = 1$. The maximum output from the hidden unit occurs when the distance measure is 0. When the input is 3 standard deviations from the mean μ, the center of the function, the output of the hidden unit is practically nil.

Radial basis function networks with more than one input unit have more parameters for each hidden node; e.g.,. if there are two input units, then the basis function for each hidden unit j needs two location parameters, μ_{1j} and μ_{2j}, for the center, and, optionally, two parameters, σ_{1j} and σ_{2j}, for variability. The dimension of the centers for each of the hidden units matches the dimension of the input vector.

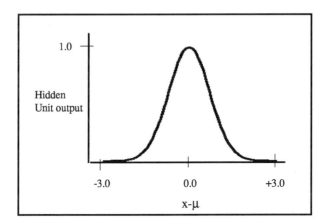

Figure 4.3 *Output from a Gaussian radial basis function for a single input value x.*

So, it can be seen that, computationally, a radial basis function network has three separate actions:

1) compute the distance from the input layer (usually a vector) to each of the hidden units (each of which represents an ellipsoid region in data space);

2) compute the output (a function of the distance measure in step 1) from each hidden unit;

3) and, finally, use the outputs of the hidden units to compute the network output with a simple weighted sum function.

The only difficult part is finding the values for μ and σ for each hidden unit, and the weights between the hidden and output layers, i.e., training the network. This will be discussed later, in Chapter 5. At this point, it is sufficient to say that training radial basis function networks is considerably faster than training multilayer perceptrons. On the other hand, once trained, the feed-forward process for multilayer perceptrons is faster than for radial basis function networks.

4.1.2 Applications

Radial basis function networks can be applied to the same types of function approximation and classification problems as multilayer perceptrons; however, the most common use of radial basis function networks seems to be for classification.

Hypothetical Example

Figure 4.4 illustrates a radial basis function solution to the hypothetical depression classification problem. The algorithm that implemented the radial basis function application determined that seven cluster sites were needed for this problem. A constant variability term, $\sigma^2 = .1$, was used for each hidden unit. Shown in the diagram are the two central parameters (because there are two input units) for each of the seven Gaussian functions.

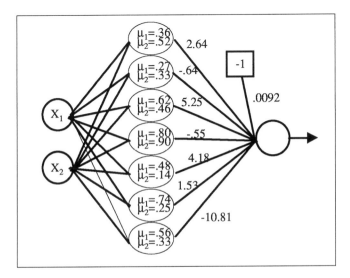

Figure 4.4 *Radial basis function solution to the depression classification problem.*

To illustrate the calculations for a test patient, assume that the score on the first measure is .48 and the second measure .14. The Euclidean distance from this vector

$$x = \begin{bmatrix} .48 \\ .14 \end{bmatrix} \quad \text{and the cluster mean given by the first hidden unit } \mu = \begin{bmatrix} .36 \\ .52 \end{bmatrix} \text{ is}$$

$$D = \sqrt{(.48 - .36)^2 + (.14 - .52)^2} = .398$$

From equation 4.2 the output of the first hidden unit in response to this input vector is then

$$h_1(D_1^2) = e^{-(.398^2/2(.1))} = .453 \quad .$$

The same calculation must be repeated for each of the remaining hidden units. Once the output of each hidden unit is obtained, the final output is the simple weighted sum of all the inputs to the output node:

$$(.453)(2.64) +(.670)(-.64)+(543)(5.25) \ldots +(-1)(.0092) = 0.0544$$

Since the output is close to 0, the patient is classified as not depressed.

Protein Folding Example

 Despite the fact that the neural network literature increasingly contains examples of radial basis function network applications, their use in genome informatics has rarely been –reported--not because the potential for applications is not there, but more likely due to a lag time between development of the technology and applications to a given field. Casidio *et al.* (1995) used a radial basis function network to optimally predict the free energy contributions due to hydrogen bonds, hydrophobic interactions and the unfolded state, with simple input measures.

4.1.3 Limitations

The same limitations that apply for multilayer perceptron networks, in general, hold for radial basis function networks. Training time is enhanced with radial basis function networks, but application is slower, due to the complexity of the calculations. Radial basis function networks require supervised training and hence are limited to those applications for which training data and answers are available. Several books are listed in the reference section with excellent descriptions of radial basis function networks and applications (Beale & Jackson, 1991; Fu, 1994; Wasserman, 1993).

4.2 Kohonen Self-organizing Maps

Perceptrons, multilayer perceptrons and radial basis function networks require supervised training with data for which the answers are known. Some applications require the automatic clustering of data, data for which the clusters and clustering criteria are not known. One of the best known architectures for such problems is the Kohonen self-organizing map (SOM), named after its inventor, Teuvo Kohonen (Kohonen, 1997). In this section the rationale behind such networks is described.

4.2.1 Background

Network diagrams and symbols for the self-organizing map networks are quite different from those used for perceptrons, multilayer perceptrons and radial basis function networks. Input units are the same, but there are no hidden or output layers or artificial neurons. Essentially, a self-organizing map (often called a Kohonen map) is a grid of discrete points, each of which is connected to all of the input units (see Figure 4.5). Each connection between an input unit and grid unit is associated with a weight, or measure of strength (as for multilayer perceptrons). Typically, self-organizing maps are shown without the input vector.

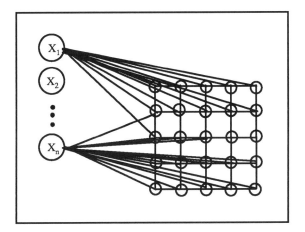

Figure 4.5 *Kohonen self-organizing map diagram.*
(Not all connections are shown).

Input vectors are mapped to one of the grid points in a trained map. A grid point and several surrounding points (a neighborhood) may each represent a different object or cluster of objects. Operation and training of a self-organizing map are similar to that for the first stage of a radial basis function network: a distance metric (Euclidean distance) is computed for each grid point for a given input vector. If the distance is small, then the vector is said to belong to the cluster (or group or class) signified by the grid point. If the distance is large then it is unlikely that the vector belongs to that cluster.

For a trivial example, assume that a self-organizing map has been trained and two groups identified, based on the absence or presence of four input features. Figure 4.6 shows the connections between the input vector and two grid points (often called nodes) that are associated with their respective groups.

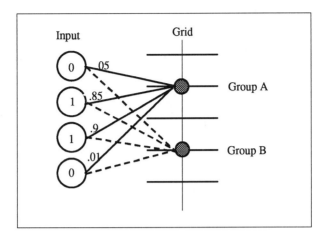

Figure 4.6 *Self-organizing map, simple example.*
Weights from input to Group B node (dotted lines, values not shown) are (0.5,0.6,0.7,0.8), respectively.

The Euclidean distance from input vector (0,1,1,0) and the weights to Group A node is

$$D_A = \sqrt{(0 - 0.05)^2 + (1 - .85)^2 + (1 - .9)^2 + (0 - .01)^2} = .187$$

For Group B node the Euclidean distance is $D_B = 1.068$. So, the new object (whatever it is) with this input vector is more likely to be in Group A than in Group B, i.e., the new object has been *assigned* to Group A.

4.2.2 Applications

Amino Acid Grouping Example

Table 6.1 (in Chapter 6) shows physical characteristics of all 20 amino acids. A self-organizing map was trained with four of these characteristics (features): mass in Daltons, K&D hydrophobicity, surface area and turn propensity. Mass was scaled to the unit interval. No other information was presented to the self-organizing map. A 10 x 10 grid was used. The results are shown in Figure 4.7, after only 50 training iterations; each intersection of lines in the table represents a grid point. Each amino acid was mapped into one of the 100 grid points.

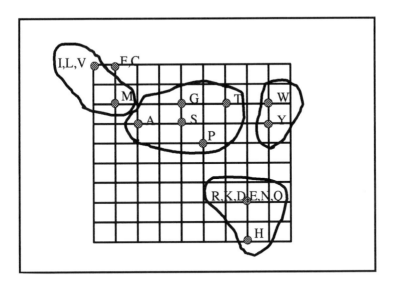

Figure 4.7 *Self-organizing map results after 50 training iterations with four amino acid features.*

It can be seen that the groupings found automatically by the self-organizing map resemble the exchange groups as found by elaborate examinations of sequence conservation. The traditional exchange groups are {H,R,K}, {D,E,N,Q} (these first two groups are often combined into one group), {S,T,P,A,G}, {M,I,L,V}, {F,W,Y}, {C}. Only the F and C amino acids are out of place according to these groups. This is a remarkable achievement for a simple network, working unsupervised, with only four amino acid features as input data. This simple application illustrates the potential power of self-organizing maps.

Protein Sequence Example

An interesting example of using Kohonen maps for the analysis of protein sequences is given in a journal article by Hanke (Hanke & Reich, 1996). In this application, a trained Kohonen map network was used to identify protein families, aligned sequences or segments of similar secondary structure, in a highly visual manner.

Other Recent Examples

Recent applications of Kohonen self-organizing maps include construction of phylogenetic trees from sequence information (Wang *et al.*, 1998); predicting cleavage sites in protein (Cai *et al.*, 1998a); prediction of beta-turns (Cai *et al.*, 1998b);

classification of structural motifs (Schuchhardt *et al.*, 1996); and classification of protein sequences (Ferran *et al.*, 1994).

4.2.3 Limitations

Self-organizing maps are, like most neural network applications, limited by the quality of the data that is used to train them. If the training data is not representative of the whole set of data to which a network is expected to apply, then the clusters found in training may not be representative. This is especially critical with small sets of training data.

4.4 References

Beale, R. & Jackson, T. (1991). *Neural Computing: An Introduction*. Adam Hilger, Bristol.

Cai, Y. D., Yu, H. & Chou, K. C. (1998a). Artificial neural network method for predicting HIV protease cleavage sites in protein. *J. Protein Chem* **17**, 607-15.

Cai, Y. D., Yu, H. & Chou, K. C. (1998b). Prediction of beta-turns. *J. Protein Chem* **17**, 363-76.

Casadio, R., Compiani, M., Fariselli, P. & Vivarelli, F. (1995). Predicting free energy contributions to the conformational stability of folded proteins from the residue sequence with radial basis function networks. *Intelligent Systems for Molecular Biology* **3**, 81-8.

Ferran, E. A., Pflugfelder, B. & Ferrara, P. (1994). Self-organized neural maps of human protein sequences. *Protein Sci* **3**, 507-21.

Fu, L. (1994). *Neural Networks in Computer Intelligence*. McGraw-Hill, New York.

Hanke, J. & Reich, J. G. (1996). Kohonen map as a visualization tool for the analysis of protein sequences: multiple alignments, domains and segments of secondary structures. *Comput Applic Biosci* **6**, 447-54.

Kohonen, T. (1997). *Self-Organizing Maps*, second edition. Springer, Berlin.

Schuchhardt, J., Schneider, G., Reichelt, J., Schomberg, D. & Wrede, P. (1996). Local structural motifs of protein backbones are classified by self-organizing neural networks. *Protein Eng* **9**, 833-42.

Wang, H. C., Dopazo, J., de la Fraga, L. G., Zhu, Y. P. & Carazo, J. (1998). Self-organizing tree-growing network for the classification of protein sequences. *Protein Sci* **7**, 2613-22.

Wasserman, P. D. (1993). *Advanced Methods in Neural Computing*. Van Nostrand Reinhold, New York.

CHAPTER 5

Training of Neural Networks

Neural networks are models that may be used to approximate, summarize, classify, generalize or otherwise represent real situations. Before models can be used, however, they must be trained or 'fit' to representative data. The model parameters, e.g., number of layers, number of units in each layer and weights of the connections between them, must be determined. In ordinary statistical terms, this is called regression. There are two fundamental types of training (or learning) with neural networks: supervised and unsupervised learning. Some architectures, e.g., radial basis functions, use a combination of supervised and unsupervised learning. For supervised training, as in regression, data used for the training consists of independent variables (also called feature variables or predictor variables) and dependent variables (target values). The independent variables (input to the neural network) are used to predict the dependent variables (output from the network). In the hypothetical example in Chapter 3 (Figure 3.4), the independent variables are the two mood measures, and the dependent (target) variable is the classification *depressed* or *not depressed*. The training set was the 19 observed pairs of observations with the known classification for each pair. Unsupervised training does not have dependent (target) values supplied: the network is supposed to cluster the data automatically into meaningful sets. The training set for the Kohonen self-organizing map example (see Figure 4.7) was the four amino acid features for each of the 20 amino acids.

The fundamental idea behind training, for all neural network architectures, is this: pick a set of weights (often randomly), apply the inputs to the network and see how the network performs with this set of weights. If it doesn't perform well, then modify the weights by some algorithm (specific to each architecture) and repeat the procedure. This iterative process is continued until some pre-specified stopping criterion has been reached.

A training pass through all vectors of the input data is called an *epoch*. Iterative changes can be made to the weights with each input vector, or changes can be made after all input vectors have been processed. Typically, weights are iteratively modified by epochs.

5.1 Supervised Learning

In Part II of this book we have encountered three network architectures that require supervised learning: perceptrons, multilayer perceptrons and radial basis function networks. Training for perceptrons and multilayer perceptrons is similar. The goal of training is to find the set of parameters (number of layers, number of units within layers and the weights between layers) that minimizes the difference between the network output values and the desired target values. A typical error function to be minimized is

$$E = \sum_{i=1}^{n} (o_i - t_i)^2 \qquad (5.1)$$

where n is the number of input data vectors (patterns), o_i and t_i are the network output (for a given set of parameters and input vectors) and target values, respectively. If a network has more than one output unit in the output layer, then equation 5.1 becomes

$$E = \sum_{i=1}^{n} \sum_{j=1}^{k} (o_{ij} - t_{ij})^2 \qquad (5.2)$$

where k is the number of units in the output layer.

Although there are a number of mathematical methods used to minimize the error function as a function of the weights, methods to find optimal numbers of layers and units within layers are usually determined heuristically. Function minimization (training) methods described below assume that the number of layers and units already has been determined and the only remaining problem is to determine the set of weights between them. Some of the training methods described below, such as back-propagation, represented important advances in neural network science when they were devised and are still widely used; however, there are often much better methods of training based on research in other fields such as numerical analysis, differential equations and statistics. It is important to emphasize that any method of modifying the network parameters to reduce the error is called *learning*. The methods traditionally used by neural network practitioners, such as back-propagation, do not necessarily have to be used when other, often more familiar or faster, methods will achieve the same goals. For a comparison of neural network approaches and traditional statistical and numerical analysis approaches, see Cheng and Titterington (1994), and Sarle (1994).

5.2.1 Training Perceptrons

The principle behind training perceptrons is simple: try a set of weights. If the output from the network, in response to a given input vector, matches the target value for that input vector, then go to the next input vector, and the goal has been reached. If the output value is larger than the target value, reduce the weights and try again. If the output value is smaller than the target value, increase the weights and try again. This process is repeated for all input vectors until there is no change to be made for any of them.

Formally,

1) Initialize all weights, including threshold values, to a small random number, say, between 0 and 1;

2) Apply an input vector with components $(x_1, x_2,..x_I)$ to the network, calculate o, the output of the network for this vector. For notational simplicity, assume that there is just one input vector X with i elements, and its associated target value t;

3) Modify the weights, as necessary:

   if o = t, choose another input vector and go to step 2

   if o < t, make each weight value bigger, $w_i \Leftarrow w_i + x_i$

   if o > t, make each weight smaller, $w_i \Leftarrow w_i - x_i$

4) Repeat 2 and 3 for the new weights, until there are no more changes to be made for each input vector. The symbol "$\Leftarrow$" means replace the old value with a new value.

This is called the perceptron learning rule and it has been proven to converge to a solution, for linearly separable problems, in a finite number of iterations. The weight adjustment rule can be restated as

new weight = old weight + delta*x_i

where in this case delta is (t-o) and '*' means multiplication; this is called the delta rule. In words, the change in the weight to an output unit is proportional to the error (delta) and the input to the unit.

To illustrate this procedure, consider the partial table of input and target values (Table 5.1) for the hypothetical example discussed in Chapter 3 (data shown in Figure 3.1).

Table 5.1 *Sample data for training a perceptron.*

Input Vector Number	Measure 1	Measure 2	Target (known diagnosis) 0 = not depressed, 1= depressed
1	.2	.85	1
2	.37	.83	1
3	.62	.90	1
.	.	.	.
.	.	.	.
.	.	.	.
18	.74	.25	0

The perceptron network for this problem is shown in Figure 5.1 below. Input from x_0 , the bias unit, is always -1. The length of the input vector is 2, and there are 18 input vectors in the training set.

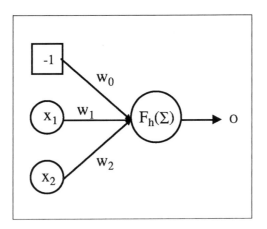

Figure 5.1 *Simple perceptron network for training example.*

Assume that the weights were randomly assigned values $w_0 = .3$, $w_1 = .2$ and $w_2 = .1$,

For the first input vector ($x_0 = -1$, $x_1 = .2$, $x_2 = .85$), the output of the network is less than the target value (the weighted sum of the inputs to the output unit is less than 0, so the output is mapped by the Heaviside function to 0):

$$o = 0 \text{ and } t = 1.$$

Modify the input weights to make them bigger:

$w_0 \Leftarrow w_0 + x_0 = .3 + (-1) = -.7$

$w_1 \Leftarrow w_1 + x_1 = .2 + .2 = .4$

$w_2 \Leftarrow w_2 + x_2 = .1 + .85 = .95$

Applying the input vector again, with these new weights, the output of the network is 1,

$$o = 1 \text{ and } t = 1.$$

So iterative modification of the weights is stopped **for this input vector**. The next input vector is chosen and the process repeated. This process is repeated until all the input vectors produce the same output value as their target values.

There are modifications to the perceptron learning rule to help effect faster convergence. The Widrow-Hoff delta rule (Widrow & Hoff, 1960) multiplies the delta term by a number less than 1, called the learning rate, η. This effectively causes smaller changes to be made at each step. There are heuristic rules to decrease η as training time increases; the idea is that big changes may be taken at first and as the final solution is approached, smaller changes may be desired.

The perceptron delta rule can be seen as one technique to achieve the function minimization of equation 5.1. There are more effective methods that will be discussed in later sections.

5.2.2 Multilayer Perceptrons

One of the early problems with multilayer perceptrons was that it was not clear how to train them. The perceptron training rule doesn't apply directly to networks with hidden layers. Fortunately, Rumelhart and others (Rumelhart *et al.*, 1986) devised an intuitive method that quickly became adopted and revolutionized the field of artificial neural networks. The method is called *back-propagation* because it computes the error term as described above and propagates the error backward through the network so that weights to and from hidden units can be modified in a fashion similar to the delta rule for perceptrons.

The back-propagation algorithm

Since multilayer perceptrons use neurons that have differentiable functions, it was possible, using the chain rule of calculus, to derive a delta rule for training similar in form and function to that for perceptrons. The result of this clever mathematics is a powerful and relatively efficient iterative method for multilayer perceptrons. The rule for changing weights into a neuron unit becomes

New weight = old weight + delta*(input into the unit)

This has the same form as the perceptron delta rule but the terms are a little more complicated.

For an output unit k, delta is given the symbol δ_k,

$$\delta_k = o_k (1 - o_k)(t_k - o_k)$$

For a hidden unit j,

$$\delta_j = o_j(1 - o_j)\sum_k \delta_k w_{jk}$$

Then the rule for changing weights becomes

$$w_{ij} \Leftarrow w_{ij} + \eta\delta_j o_j \qquad (5.3)$$

where η is the rate modifier (learning rate) as before. In words, modify the weight between unit i and unit j with a change proportional to the error from unit j and the output of unit i into unit j. Note that δ is proportional to the error (target value – observed output value). The messy looking equation for δ for hidden units is a way of distributing the error observed at an output unit back to the hidden units to which it is connected (shown graphically in Figure 5.2). Detailed derivations of the training rule for multilayer perceptrons are included in a number of textbooks (Beale & Jackson, 1991; Hinton, 1992).

Although back-propagation is a powerful method of training multilayer perceptrons, it is not without problems. Some problems may require thousands or tens of thousands of iterations before convergence is achieved, taking days to complete even on high-speed computers. Numerous variations of back-propagation have been devised to speed up the process. Most of them work by adjusting the learning rate to optimize the size of the changes made at each iteration. It is not necessary to use back-propagation at all; standard nonlinear regression algorithms, developed for use in other fields of research, can perform as effectively and sometimes better. Any method of minimizing the error function in Equation 5.1 by manipulating the weight parameters in the network will do (Cheng & Titterington, 1994; Sarle, 1994). Algorithms devised for the solution of *stiff* differential equations -- differential equations that have widely varying time constants -- have been successfully applied to multilayer perceptron training.

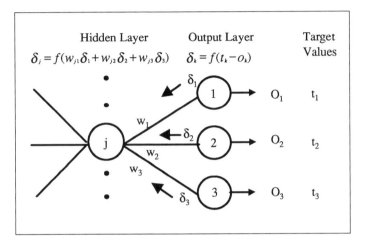

Figure 5.2 Illustration of back-propagation of errors.

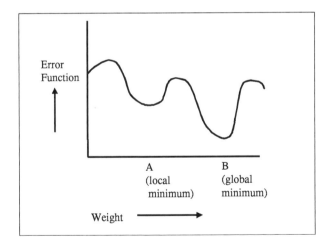

Figure 5.3 Global and local minima.

One problem faced by all methods of solution is that of finding the *global minimum* of an error function. There is no guarantee that a solution found will be the best solution. To illustrate, Figure 5.3 shows a function in two dimensions. In this figure a numerical search for minima that starts from the left side of the graph will probably get stuck in a local minimum (point A); the error function doesn't change over a range of weight values, whereas starting from the right side will probably lead to the global minimum (point B). The common approach to this problem is to run the minimization algorithm (back-propagation or whatever) from several different randomly distributed starting points and choose the best solution from the set of solutions.

Another method used to help avoid local minima and to help increase the convergence rate is the addition of a *momentum* term to Equation 5.3. This term allows the weight change at any iteration to be a function of the weight change for the previous iteration. This prevents abrupt changes in the weight modification process that potentially may lead to getting trapped in local minima. Equation 5.3 is modified to

$$w_{ij} \Leftarrow w_{ij} + \eta\delta_j o_j + \alpha\Delta$$

where α a constant ($0<\alpha<1$) called the momentum factor and Δ is the magnitude of the weight adjustment ($\eta\delta_j o_j$) made at the previous iteration.

5.2.3 Radial Basis Functions

Radial basis function networks and multilayer perceptrons have similar functions but their training algorithms are dramatically different. Training radial basis function networks proceeds in two steps. First the hidden layer parameters are determined as a function of the input data and then the weights between the hidden and output layers are determined from the output of the hidden layer and the target data.

The hidden layer parameters to be determined are the parameters of hyperellipsoids that partition the input data into discrete clusters or regions. The parameters locate the centers (i.e., the means) of each ellipsoid region's basis function and describe the extent or spread of the region (i.e., the variance or standard deviations). There are many ways of doing this. One is to use random samples of the input data as the cluster centers and add or subtract clusters as needed to best represent the data. Perhaps the most common method is called the K-means algorithm (Kohonen, 1997; Linde *et al.*, 1980):

1) Randomly choose K vectors from the input data set to be the centers of K basis functions.

2) For each vector in the input data set compute the Euclidean distance (Equation 4.1) to each of the K basis function centers.

3) Determine the closest basis function center for each input data vector.

4) For all the input vectors grouped around basis functions, compute the mean.

5) Use these grouped means as the new mean values for the K basis functions.

6) Repeat this process until there is no further significant change to the basis function centers.

This process assumes advance knowledge of how many basis functions (or data clusters) will be required to appropriately partition the data space. There are numerous heuristic methods of addressing this issue. Since training radial basis function networks is rapid, it is easy to start with a small number of centers and iteratively increase the number until no further benefit is noticed. Note that the dimension of the basis function means is the same as the dimension of the input data vectors.

The second part of training radial basis function networks assumes that the number of basis functions, i.e., the number of hidden units, and their center and variability parameters have been determined. Then all that remains is to find the linear combination of weights that produce the desired output (target) values for each input vector. Since this is a linear problem, convergence is guaranteed and computation proceeds rapidly. This task can be accomplished with an iterative technique based on the perceptron training rule, or with various other numerical techniques. Technically, the problem is a matrix inversion problem:

$$T = XW$$

where T is the target vector, W is the to-be-determined weight vector and X is the matrix of output values from each hidden unit in response to the input data (calculated from the basis functions, Equation 4.2). The matrix is usually not square, so a pseudoinverse may be used to give a minimum least square solution. Simple linear regression software tools can be used to accomplish the solution.

It can thus be seen that training radial basis function networks uses both supervised and unsupervised learning: determining basis function parameters is unsupervised and solution of the linear equations is supervised.

5.2.4 Supervised Training Issues

In addition to the described training algorithms for supervised learning, several practical problems need to be addressed during training. One serious problem with multilayer perceptrons is that overtraining can occur. The problem of when to stop training can be addressed by using an independent set of data called a validation set.

Overtraining

Overtraining means that the trained multilayer perceptron performance matches the training data so precisely that *generalization* cannot occur. The point of training neural networks is to use them for data they have not been trained on, i.e., the learning can be generalized to apply to new situations (mathematically, the trained network can interpolate between the training data points). A *smoothed* rather than an exact fit of the data is usually desired. This is illustrated in Figure 5.4. It is obvious from the figure that interpolation between points with the overtrained model will give meaningless answers. Generalization is not possible.

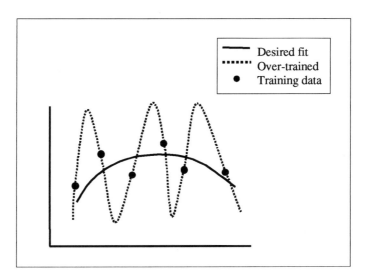

Figure 5.4 Effects of overtraining.

Validation Data

To prevent overtraining, a validation set of data can be used. This is a set of data (it can be a random subset of the whole training set of data) used not to train the network but to evaluate the results of training. The principle is this: as training proceeds, the training error function (difference between the output and target values) will decrease and the results of applying a validation set will improve (as measured by the difference between validation target and output data, or as percentage of data points correctly predicted).

However, after some point, further iterations will likely not improve performance on the validation set. At this point, training should be stopped. Rigid numeric criteria can be applied as a stopping rule, or this can be determined visually. Another approach to the overtraining (or overfitting) problem is to reduce the complexity of a network to a minimum. Using more hidden units than is necessary is analogous to over-parameterization in statistical modeling. Noise in the input data may be learned by a network, thereby reducing its ability to generalize.

Test Data

It is not particularly useful to report results of training as number of points correctly classified, using training data as the only measure of performance. Of course, results on the training data will be good if the model is reasonable; what is desired is good performance on data not used in the training, i.e., generalization. For this a separate set of data, called the test data set, is used. These terms--training data, validation data and test data--have specific meanings in the neural network literature.

Sampling Data

If data is hard to obtain, it may not be feasible to have training, validation and test data. One way of handling this is to take a sample of data for training, and test the results of applying the remaining data to the network. An extreme example of this method is called the *leave-one-out* approach: all data but one selected data point (vector) is used in the training and performance evaluated on the remaining selected data point. This can be repeated many times randomly, or each data point can be systematically left out, and summary performance measures calculated. A more reasonable approach, if sufficient data is available, is to take a random sample for testing and train with the remaining data. This technique, called *bootstrapping*, is done repeatedly and summary statistics are calculated

Data Preparation

Although not always necessary, scaling the input data to a neural network before training is advisable. There are no theoretical reasons for doing so, but for numerical reasons-- e.g., ill-conditioning, computer over-flow/under-flow, etc.--it may help speed convergence. Typical methods include scaling the input data so that the smallest data item is 0 and the largest 1. Another method (called standardization) subtracts the mean value from each data point and divides by the standard deviation, which produces smaller numbers centered around 0. The disadvantage to scaling input data is that the data is no longer in familiar units and ranges, and results of training and testing may have to be scaled back to facilitate understanding.

5.3 Unsupervised Learning

The only completely unsupervised learning architecture discussed in this book is the Kohonen self-organizing map. The Kohonen architecture assumes that an input layer is mapped to a grid, with each input unit connected to all points on the grid, as shown in Figure 4.5. Weights between the input layer and the grid are adjusted so that each input vector presented is represented by some point on the grid. If *learning* takes place, then similar inputs are mapped to the same grid points or points close by (the neighborhood). The exact algorithm is as follows:

1) Initialize the weights to small random numbers, usually between 0 and 1. The weight between input unit i and grid node j is w_{ij}.

2) Define a neighborhood radius, R, to be some large number of grids around a given point (for example a neighborhood might be a grid point and all surrounding grid points with a radius of 10 grid points or perhaps a percentage of the total grid size).

3) For a given input vector, compute the distance to each grid point, where distance is defined, for example, as the Euclidean distance (Equation 4.1).

4) Find the grid point k with the smallest distance measure.

5) Strengthen the connection between the input layer and the grid point with the minimum distance (and all points in its neighborhood):

   $w_{ij} \Leftarrow w_{ij} + \eta\delta$ where η is a learning rate modifier $(0<\eta<1)$ as before
   and $\delta = (x_{ij}-w_{ij})$ for all ij in the neighborhood of k. Note the similarity to the delta rule for other architectures.

6) Repeat steps 2 through 5 for all the patterns in the input set until some stopping criterion is met, e.g., number of iterations, or δ is less than some pre-specified small number for all input patterns. Note that η and the neighborhood size decrease as the number of training iterations increase.

As with the other training algorithms this simple technique works amazingly well. Variations on the algorithm abound. Also, as for the other algorithms, there are many model parameters that need to be determined experimentally, such as the number of grid points, the size of the neighborhoods, optimal values for the learning rate and optimal methods of reducing the learning rate and neighborhood size.

5.4 Software for Training Neural Networks

The learning algorithms presented in this chapter are relatively simple and many experimenters write their own software for implementing them. This is especially true in many genome informatics applications, where so much pre-processing of data is necessary before   applying it to neural networks. General purpose software, like SAS/IML® SAS/STAT® or numerical methods libraries, such as Visual Numerics IMSL® FORTRAN and C development tools can be used to implement many of the techniques. However, a considerable number of commercial software packages specific to neural network applications are now available that make it easier for non-programmers to use neural networks for their specific applications. Free software of variable quality is also available over the Internet.

5.5 References

Beale, R. & Jackson, T. (1991). *Neural Computing: An Introduction*. Adam Hilger, Bristol.

Cheng, B. & Titterington, D. M. (1994). Neural Networks: a review from a statistical perspective. *Stat Sci* **9**, 2-54.

Hinton, G. E. (1992). How neural networks learn from experience. *Sci Am* **267**, 144-51.

Kohonen, T. (1997). *Self-organizing maps*, 2nd ed. Springer, Berlin.

Linde, Y., Buzo, A. & Gray, R. (1980). *IEEE Trans Comm* **COM-28**, 84.

Rumelhart, D. E., Hinton, G. E. & Williams, R. J. (1986). Learning representations by back-propagating errors. *Nature (London)* **323**, 533-6.

Sarle, W. S. (1994). Neural networks and statistical models. In *Nineteenth Annual SAS Users Group International Conference*.

Widrow, B. & Hoff, M. E. Jr. (1960). Adaptive switching circuits. In *IRE WESCON*, Convention Record, pp. 96-104.

PART III

Genome Informatics Applications

Artificial neural networks are now widely used to solve various problems in genome informatics and molecular sequence analysis. Part III provides an in-depth discussion of special system designs and considerations for building neural networks for genome informatics applications (chapters 6-8), and broad reviews of state-of-the-art methods and their evaluations (chapters 9-11).

The special design issues discussed are feature representation (chapter 6), data encoding (chapter 7), and neural networks (chapter 8). Chapter 6 summarizes important amino acid and protein features, including physicochemical and structural properties, local and global protein context and evolutionary information. Also covered are methods for feature representation, which are crucial for information extraction and data encoding. Chapter 7 examines methods for input and output encoding, as well as feature analysis for input combination and trimming. Chapter 8 analyzes many aspects of neural network design, including choices of network architecture, learning paradigm and other network parameters. The chapter also covers topics on data integrity and system performance evaluation.

The applications covered are gene recognition and nucleic acid sequence analysis (chapter 9), protein structure prediction (chapter 10), and protein family classification and sequence analysis (chapter 11). Chapter 9 describes neural network and hybrid methods used in DNA coding region recognition and gene identification, transcription and translational signal recognition, feature analysis and extraction, and nucleic acid sequence classification. Chapter 10 provides a review of neural network systems in protein secondary/tertiary structure prediction, structural classification, and structural feature analyses. Chapter 11 reviews neural network and related systems in the predictions of signal sequences and other motif regions or sites, and in protein family identification and sequence clustering.

CHAPTER 6

Design Issues – Feature Presentation

6.1 Overview of Design Issues

Applications of neural networks to genome informatics share several similar design issues that affect system performance and integrity. The major issues include the input/output encoding (i.e., pre-processing and post-processing), the neural network design, the training/prediction data compilation and system evaluation mechanism (Figure 6.1).

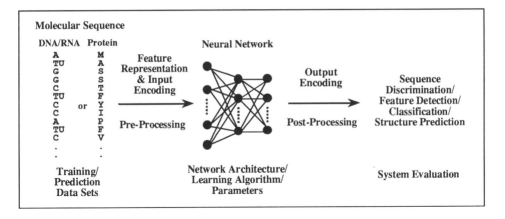

Figure 6.1 Design issues of neural network applications for genome informatics.

The first step in developing a neural network system is problem definition, i.e., to determine the desired input/output mapping for the task. The design of a complete system then involves the choices for a pre-processor and a post-processor, in addition to the neural network model itself. Data pre-processing is an integral and crucial part of any neural network system design. Pre-processing involves both feature representation (knowledge extraction) and input encoding. The selection of appropriate feature representation and encoding methods is probably one of the most significant factors determining the performance of the final system because it determines what information is presented to the neural networks. Feature representation is critical to molecular sequence analysis for two reasons: (i) this is how *a priori* knowledge about molecular structures and functions can be utilized to improve accuracy, and (ii) it allows the extraction of salient features that are characteristics of given sequences.

As domain knowledge is crucial to system design, this chapter will present an in-depth discussion of the current state of knowledge and how it can best be represented. We start by describing the twenty amino acid residues (6.2), their physicochemical properties and structural features (6.3). This provides a convenient mechanism for mapping a linear protein sequence into a linear feature space. It is well recognized that *most amino acids play more than one structural or functional role*. Some roles could be distinguished by context in the sequence. The context may be local, covering only neighboring residues; or it may be global, containing long-range spatial relationship. Other roles could be distinguished by patterns of amino acid replacement in homologous proteins through evolution. Therefore, it is also important to discuss the protein contextual (6.4) and evolutionary (6.5) features. Finally, the features may be represented in different ways (6.6) to facilitate information extraction and data encoding (chapter 7). The discussion in this chapter is also relevant and applicable to other non-neural network approaches for mapping, classification and clustering.

6.2 Amino Acid Residues

Natural proteins are built of various combinations of 20 amino acids, and have precise lengths and exact sequences of amino acids. Indeed, only the differences in length and sequence distinguish one protein from any other and make possible a diversity of structures and functions. A common approach to the molecular sequence analysis problem is to consider a protein or nucleic acid sequence a string of letters in a defined alphabet set. However, the letters used in the biological alphabets are more than just letters. They have biological meaning that should be explored in order to develop an intelligent system for understanding the codes connecting sequences to their structures and functions.

The secret to the functional diversity of proteins lies partly in the chemical diversity of the amino acids but primarily in the diversity of the three-dimensional structures that these building blocks can form, simply by being linked in different sequences. At the heart of the determination of structure by sequence lie the distinctive characteristics of each of the 20 different amino acids. Table 6.1 shows some general properties of the amino acids along with their three- and one-letter abbreviations.

Glycine, Pro and Cys have the most distinctive geometries and features. Glycine is the smallest amino acid, with only a hydrogen atom for a side chain. This allows Gly to have greater conformation flexibility and fit in where other residues would be too bulky. Proline lacks a free amide hydrogen group to prevent main chain hydrogen bonding, and has a high additional constraint on the backbone relative to any other amino acid. The thiol group of Cys residues is the most reactive of any amino acid side chain. Cysteine is notable both for its ability to serve as ligands to a variety of metals and prosthetic groups, and also for its ability to form disulfides. There are three very different types of Cys, then: (1) free SH groups, (2) ligand SH groups, and (3) S-S disulfides (cystines).

Alanine is the default amino acid with no long side chain, chemical reactivity or unusual conformational properties. Valine, Ileu, Leu, and Met are the aliphatic residues that fit together the hydrophobic interiors of proteins. They do not interact favorably with water; instead they interact more favorably with each other and with other nonpolar atoms. Serine and Thr are hydroxyl residues. Both are chemically active, so that they have important roles as active site residues, can be phosphorylated, occur readily in acetylation with acetyl chloride, and can be carbohydrate attachment sites. Aspartate and Glu are acidic residues, Asn and Gln are the amide forms of Asp and Glu, whereas Lys and Arg are basic residues. Histidine, Phe, Tyr, and Trp are the aromatic rings. Histidine has a charge group that is unique among amino acids in having a pK near neutrality, thereby performing as an active-site residue or as a controllable element in conformational changes. Phenylalanine, Tyr, and Trp are major constituents of the hydrophobic cores inside proteins.

6.3 Amino Acid Physicochemical and Structural Features

A large number of chemical and physical properties, manifest in the amino acid side chains, have been thoroughly examined by many investigators. Attempts have been made to correlate these properties with their relatedness among protein sequences. What is most relevant is how these side chains interact with the backbone and with one another and what roles they each play within particular types of secondary and tertiary structure. The parametric description of residue environments with the help of solvent accessibility, secondary structure, backbone torsion angles, pairwise residue-residue distances, or Cα positions is the comparison between amino acid types at protein sequence positions and residue locations in structural templates. A recent review has evaluated and quantified the extent to which the amino acid type-specific distributions of commonly used environment parameters discriminate with respect to the 20 amino acid types (Sunyaev *et al.*, 1998). Some of the important amino acid properties and residue environments are discussed below.

Hydrophobicity

Hydrophobicity is believed to play a major role in organizing the self-assembly of protein molecules. Patterns of hydrophobic versus hydrophilic side chains are very in predicting secondary and tertiary structure simply by virtue of their preferential occurrence on the inside versus outside of various structural features. Those hydrophobicity patterns provide a rough translation of the three-dimensional structural features into the one-dimensional domain of the amino acid sequence. A hydrophobicity scale assigns a hydrophobicity value to each of the 20 amino acids. Many hydrophobicity scales have been developed over the last decades on the basis of solubility measurements of the amino acids in different solvents, vapor pressures of side chain analogues, analysis of side-chain distributions with soluble proteins and the theoretical energy calculations. For example, the Kyte and Doolittle (1992) scale was based on experimental data (shown in Table 6.1), the Engelman (Engelman *et al.* 1986) scale was calibrated specifically for transmembrane helices, and the Eisenberg (Eisenberg *et al.*, 1989) scale used normalized consensus values synthesized from five earlier scales.

Table 6.1 *Physicochemical and structural properties of amino acid residues.*

Amino Acid Residues (3-, 1-letter code)	Chemical Property	Volume (A^3)	Mass (daltons)	HP Scale K&D	Surface Area	2D Structure Propensity		
						α-helix	β-strand	Turn
Alanine (Ala, A)	Aliphatic	67	71.09	1.8	0.74	1.41	0.72	0.82
Arginine (Arg, R)	Basic	148	156.19	-4.5	0.64	1.21	0.84	0.90
Asparagine (Asn, N)	Amide	96	114.11	-3.5	0.63	0.76	0.48	1.34
Aspartic Acid (Asp, D)	Acidic	91	115.09	-3.5	0.62	0.99	0.39	1.24
Cysteine (Cys, C)	Reactive	86	103.15	2.5	0.91	0.66	1.40	0.54
Glutamine (Gln, Q)	Amide	114	128.14	-3.5	0.62	1.27	0.98	0.84
Glutamic Acid (Glu, E)	Acidic	109	129.12	-3.5	0.62	1.59	0.52	1.01
Glycine (Gly, G)	Small	48	57.05	-0.4	0.72	0.43	0.58	1.77
Histidine (His, H)	Aromatic	118	137.14	-3.2	0.78	1.05	0.80	0.81
Isoleucine (Ile, I)	Aliphatic	124	113.16	4.5	0.88	1.09	1.67	0.47
Leucine (Leu, L)	Aliphatic	124	113.16	3.8	0.85	1.34	1.22	0.57
Lysine (Lys, K)	Basic	135	128.17	-3.9	0.52	1.23	0.69	1.07
Methionine (Met, M)	Aliphatic	124	131.19	1.9	0.85	1.30	1.14	0.52
Phenylalanine (Phe, F)	Aromatic	135	147.18	2.8	0.88	1.16	1.33	0.59
Proline (Pro, P)	Cyclic Imino	90	97.12	-1.6	0.64	0.34	0.31	1.32
Serine (Ser, S)	Hydroxyl	73	87.08	-0.8	0.66	0.57	0.96	1.22
Threonine (Thr, T)	Hydroxyl	93	101.11	-0.7	0.70	0.76	1.17	0.90
Tryptophan (Trp, W)	Aromatic	163	186.21	-0.9	0.85	1.02	1.35	0.65
Tyrosine (Tyr, Y)	Aromatic	141	163.18	-1.3	0.76	0.74	1.45	0.76
Valine (Val, V)	Aliphatic	105	99.14	4.2	0.86	0.90	1.87	0.41

Volume = Volume enclosed by van der Waals radii; Mass = molecular weight of nonionized amino acid minus that of water; both adopted from Creighton (1993); HP scale = degree of hydrophobicity of amino acid side chains, based on Kyte & Doolittle (1982); Surface Area = mean fraction buried, based on Rose *et al.* (1985); and Secondary structure propensity = the normalized frequencies for each conformation, adopted from Creighton (1993), is the fraction of residues of each amino acid that occurred in that conformation, divided by this fraction for all residues.

Surface Area

Buried surface area is often used as a measure of the contribution to protein folding from the hydrophobic effect. Quantitatively, the surface buried upon folding is reckoned as the difference in area between the native and unfolded states. An extensive survey of the distribution of accessible areas for each residue type has been provided by Rose *et al.* (1985) based on proteins of known structure. This characteristic quantity, the average area buried, is correlated with residue hydrophobicity. Generally, in folded structures, the amino acid residues Phe, Leu, Ile, and Met tend to be fully buried, whereas the charged groups of Arg, Lys, His, Glu, and Asp residues tend to be exposed on the surface. Later, the mean area buried upon folding for every chemical group in each residue within a set of X-ray-elucidated proteins was calculated (Lesser & Rose, 1990; Creamer *et al.*, 1997).

Secondary Structure Propensity

Using simple statistical counting of the known structures, the propensities of the different amino acids for forming α-helix, β-strand and reverse turn conformations have been calculated many times (Chou & Fasman, 1978). These preferences correlate well with the chemical structure and stereochemistry of the particular amino acids. For example, Pro lacks a free NH group preventing main chain H bonding, so rarely occurs in the center of an α-helix. Some rules that emerged from these studies include the following. The majority of turns lie on the surface of the protein and are surrounded by water; therefore, it is the hydrophilic amino acids, along with Pro and Gly, that occur most often in β-turns. Proline strongly disfavors α-helix or β-sheet formation, except at their margins. The type I' and II' turns favor the presence of a Gly residue. The branched β-carbon amino acids (Thr, Val, and Ile) are relatively poor helix formers. Some of the polar amino acids, especially Asp and Asn, often form hydrogen bonds and prohibit helix formation. Serine and Thr stabilize the helix and are commonly found in transmembrane helices. The loop regions, connecting the classic secondary structures, are generally very short and quite variable in the homologous structures. A limited set of loop conformers is observed, which often includes critical Gly residues. Most loops interact with the solvent and are highly hydrophilic.

Other Properties

Some other physicochemical and structural properties include polarity, volume, bulkiness, and refractivity.

6.4 Protein Context Features and Domains

Most amino acids play more than one structural or functional role, some of which could be distinguished by context in the sequence. For instance, the three types of Cys (S-S, free-SH, and ligand-SH) have very different conformational preferences and clustering properties and can often be assigned from properties of the entire protein. There is local protein context covering only local interactions of neighboring residues in the sequence. It can to some extent determine local structures, such as α-helices. The global context containing long-range residues that introduce cooperative effects from the rest of the structure, however, is also necessary to determine protein structure, even the local levels of structure.

Since the different domains often have quite different types of tertiary structure, they need to be described and classified separately and, if possible, should be predicted separately. Multiple domains should be considered for any sequence that suggests an internal repeat or that is longer than about 200 residues (250 for α/β structures or 100 for a disulfide-rich sequence).

Hydrophobic Moment

An effective measure of local context to predict protein local structures is the hydrophobic moment. The overall amphipathicity of any periodic structure has been quantitated in the hydrophobic moment. Even if the atomic coordinates of a protein are not known, the sequence hydrophobic moment can be estimated, as in

$$\mu_H = \left[\left(\sum_{n=1}^{N} H_n \sin(\delta_n)\right)^2 + \left(\sum_{n=1}^{N} H_n \cos(\delta_n)\right)^2\right]^{1/2}$$

where H_n is the hydrophobicity of the nth residue in a sliding window of length N, and δ is the angle (in radians) at which successive side chains emerge from the central axis of a given structure (for an α-helix, $\delta \equiv 100°$; for a β-strand, $\delta \equiv 160°$; for a flat β-sheet, $\delta \equiv 180°$) (Eisenberg et al., 1984). It is shown to be equivalent to the modulus of the Fourier transform of the hydrophobicities of the structure calculated at the relevant frequency, which would be $1/w$ where w is the number of residues between residues that face the same side and direction in the putative structure. The moments can be calculated using different amino acid hydrophobicity scales.

Hydrophobicity Profile

The hydrophobicity profile is a simple way to quantify the concentration of hydrophobic residues along the linear polypeptide chain (Rose & Dworkin, 1989). The construction of the profiles depends on the choice of the hydrophobicity scale and the window size. The profile is computed by averaging the hydrophobicity scales of amino acid residues within a sliding window. For each position in the sequence a hydropathic index is calculated. The index is the mean value of the hydrophobicity of the amino acids within a window.

Amino Acid Frequency/Composition

Another frequently used global information that covers protein context is the residue frequencies. The composition is often calibrated with that from the database as in Garnier *et al.* (1996), where only observed frequencies of amino acids and amino acid pairs are used for protein secondary structure prediction.

6.5 Protein Evolutionary Features

The evolutionary features of amino acids are often represented by substitution matrices, most notably the PAM (point accepted mutation) (Dayhoff *et al.*, 1978) and BLOSUM (BLOcks SUbstitution Matrices) (Henikoff & Henikoff, 1992) series. The PAM matrix (Figure 6.2) is a frequency table representing substitution rates for closely related proteins at the particular *evolutionary distance* represented by multiple sequence alignments. The PAM matrix has been updated using a more current protein database (Gonnet *et al.*, 1992). The BLOSUM matrix is developed based on the Blocks database containing alignments of conserved regions.

```
    A   R   N   D   C   Q   E   G   H   I   L   K   M   F   P   S   T   W   Y   V
A   2
R  -2   6
N   0   0   2
D   0  -1   2   4
C  -2  -4  -4  -5  12
Q   0   1   1   2  -5   4
E   0  -1   1   3  -5   2   4
G   1  -3   0   1  -3  -1   0   5
H  -1   2   2   1  -3   3   1  -2   6
I  -1  -2  -2  -2  -2  -2  -2  -3  -2   5
L  -2  -3  -3  -4  -6  -2  -3  -4  -2   2   6
K  -1   3   1   0  -5   1   0  -2   0  -2  -3   5
M  -1   0  -2  -3  -5  -1  -2  -3  -2   2   4   0   6
F  -3  -4  -3  -6  -4  -5  -5  -5  -2   1   2  -5   0   9
P   1   0   0  -1  -3   0  -1   0   0  -2  -3  -1  -2  -5   6
S   1   0   1   0   0  -1   0   1  -1  -1  -3   0  -2  -3   1   2
T   1  -1   0   0  -2  -1   0   0  -1   0  -2   0  -1  -3   0   1   3
W  -6   2  -4  -7  -8  -5  -7  -7  -3  -5  -2  -3  -4   0  -6  -2  -5  17
Y  -3  -4  -2  -4   0  -4  -4  -5   0  -1  -1  -4  -2   7  -5  -3  -3   0  10
V   0  -2  -2  -2  -2  -2  -2  -1  -2   4   2  -2   2  -1  -1  -1   0  -6  -2   4
    A   R   N   D   C   Q   E   G   H   I   L   K   M   F   P   S   T   W   Y   V
```

***Figure 6.2** PAM250 amino acid substitution matrix.*

For fold recognition, the amino acid substitution matrix can be replaced by a 3D-1D substitution matrix (scoring table) computed from a database of known structures and sets of sequence-structure alignments (Bowie *et al.*, 1991). Recently, the matrix has been further extended to include predicted secondary structure of the sequence using a H3P2 matrix (Rice & Eisenberg, 1997). The 3D-1D substitution matrix allows one to match the amino acid residues (and/or their secondary structure prediction) with structural (environmental) classes.

Although protein evolutionary features captured in the substitution matrix are compiled from a large database of protein sequences or structures, a sequence profile can be used to describe position-specific and family-specific evolutionary features (Gribskov *et al.*, 1987). The profile is a two-dimensional weight matrix in which the rows correspond to aligned positions in a group of sequences, and the columns correspond to each of the 20 amino acid residues, as illustrated in Figure 6.3 (adopted from Gribskov & Veretnik, 1996). The sequence profile, thus, represents the conservation weight of each aligned position along the sequence string.

Alignment	Cons	A	C	D	E	F	G	...	S	T	V	W	Y	Gap	Len
. . D V .	V	-3	-37	33	21	-40	-7	...	-9	-3	**46**	-74	-36	22	22
T . D I .	T	11	-29	10	6	-39	16	...	12	**25**	-3	-57	-35	22	22
...															
I V V V I	V	-44	-89	-95	-83	61	-85	...	-84	-45	**103**	-150	-77	100	100
L M L L M	L	-48	-105	-92	-71	-9	-82	...	-70	-50	0	-126	-46	100	100
D D D D D	D	-76	-59	**124**	-13	-60	-74	...	-69	-78	-101	-72	-39	100	100
E E E E E	E	-83	-57	-14	**122**	-60	-79	...	-79	-86	-108	-69	-120	100	100
A A A G A	A	**99**	-111	-76	-75	-128	-53	...	-45	-45	-80	-141	-125	100	100
D D D D D	D	-76	-59	**124**	-13	-60	-74	...	-69	-78	-101	-72	-39	100	100

Figure 6.3 *A sequence profile representing family-specific evolutionary features.*
Cons = the consensus sequence representing the highest scoring column in each row;
Gap and Len = gap opening and gap extension penalties.

6.6 Feature Representation

The amino acid and protein features can be represented in different ways to maximize information extraction. They may be represented as real-numbered measurements in a continuous scale (such as mass or hydrophobicity scales in Table 6.1), or as vectors of distances or frequencies (such as PAM matrix and sequence profile in Figures 6.2 and 6.3). But they can also be conveniently categorized into classes based on these properties. This effectively reduces the original 20-letter amino acid alphabet set to an alternative alphabet set of smaller sizes and emphasizes the various properties of the molecular residues and maximizes feature extraction.

Using the hydrophobicity scale as an example, the amino acid residues can be represented by the real-numbered scalar value. They can also be classified with respect to their side chains as polar, nonpolar, or amphipathic, depending on the range of the hydrophobicity in the scale, such as in

$$\text{Group 1}: \quad \geq (H_m + sd/2);$$
$$\text{Group 2}: \quad \geq (H_m - sd/2) \ \& < (H_m + sd/2); \qquad H_m = \frac{1}{2}\sum_{i=1}^{20} H_i$$
$$\text{Group 3}: \quad < (H_m - sd/2)$$

where H_m is the mean value of the hydrophobicity over 20 amino acids, sd is the standard deviation, and H_i is the hydrophobicity value in the scale (Nakata, 1995).

Likewise, alternative alphabets can be used to represent other physicochemical and structural features of the amino acids. Multiple alphabet comparisons can expose significant regions that are conserved in some, but not in other, alphabets. Significance of the same positions in many alphabets enhances their possible functional importance. Various amino acid groups have been described by Karlin *et al.* (1989) and Nakata (1995), including those for charge and polarity, hydrophobicity, hydrophilicity and secondary structure propensity (Table 6.2). The evolutionary matrix can also be represented as membership in a group. A six-letter exchange (substitution) group alphabet derived from the PAM matrix contains information about conservative replacement in evolution and has been shown to be highly effective for protein classification (Wu *et al.*, 1992). Several other substitution groups have been similarly derived from the PAM matrix, including a 5-letter property groups and 13-letter sub-property groups (Nakata, 1995), and a 7-letter property groups (Rice & Eisenberg, 1997).

Table 6.2 *Alphabet sets for feature representation: some examples.*

Alphabet Name	Size	Features	Membership
AAIdentity	20	Sequence Identity	A,C,D,E,F,G,H,I,K,L,M,N,P,Q,R,S,T,V,W,Y
ExchangeGroup	6	Conservative Substitution	{HRK} {DENQ} {C} {STPAG} {MILV} {FYW}
ChargePolarity	4	Charge and Polarity	{HRK} {DE} {CTSGNQY} {APMLIVFW}
Hydrophobicity	3	Hydrophobicity	{DENQRK} {CSTPGHY} {AMILVFW}
Mass	3	Mass	{GASPVTC}{NDQEHILKM}{RFWY}
Structural	3	Surface Exposure	{DENQHRK} {CSTPAGWY} {MILVF}
2DPropensity	3	2D Structure Propensity	{AEQHKMLR} {CTIVFYW} {SGPDN}

Instead of the discrete membership described above (Table 6.2), sometimes it may be desirable to define overlapping memberships. This may apply to properties such as hydrophobicity scales or secondary structure propensity, or to amino acid chemical groups. The idea of fuzzy logic (Zadeh, 1965; Kosko, 1992) can be used to *fuzzify* the memberships. Each value will be viewed as a fuzzy set associate with a membership function, which can be triangular, bell-shaped, or of another form. The degree of membership can be interpreted as the degree of possibility using membership functions.

The features can also be expressed as hierarchical classes (substitution groups in a hierarchical arrangement) with or without overlapping memberships. (Taylor Venn diagram). The hierarchical amino acid classes can be developed based on physicochemical nature of side chains, as in Smith and Smith (1992) (Figure 6.4), or derived based on databases of aligned sequences (Nevill-Manning *et al.*, 1998). The classes have been applied to sequence alignment problem, where the task of alignment is to identify the smallest class (minimally inclusive class) that covers the amino acids at each aligned position.

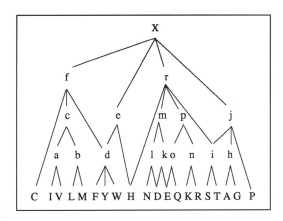

Figure 6.4 *Amino acid class hierarchy.*

6.7 References

Bowie, J. U., Luthy, R. & Eisenberg, D. (1991). A method to identify protein sequences that fold into a known three-dimensional structure. *Science* **253**, 164-70.

Chou, P. Y. & Fasman, G. D. (1978). Prediction of the secondary structure of proteins from their amino acid sequence. *Adv Enzymology Relat Areas Mol Biol*, **47**,45-148.

Creamer, T. P., Srinivasan, R. & Rose, G. D. (1997). Modeling unfolded states of proteins and peptides. II. Backbone solvent accessibility. *Biochemistry* **36**, 2832-5.

Creighton, T. E. (1993) Proteins: Structures and Molecular Properties, second ed. W.H. Freeman and Company, New York.

Dayhoff, M. O., Schwartz, R. M. & Orcutt, B. C. (1978). A model of evolutionary change in proteins. Matrices for detecting distant relationship. In *Atlas of Protein Sequence and*

Structure, vol. 15, suppl 3 (ed. Dayhoff, M. O.), pp. 345-58. National Biomedical Research Foundation, Washington, D.C.

Eisenberg, D., Weiss, R. M. & Terwilliger, T. C. (1984). The hydrophobic moment detects periodicity in protein hydrophobicity. *Proc Natl Acad Sci U S A* **81**, 140-4.

Engelman, D. M., Steitz, T. A. & Goldman, A. (1986). Identifying nonpolar transbilayer helices in amino acid sequences of membrane proteins. *Annu Rev Biophys Biophys Chem* 1986;**15**:321-53.

Garnier, J., Gibrat, J. F. & Robson, B. (1996). GOR method for predicting protein secondary structure from amino acid sequence. *Methods Enzymol* **266**, 540-53.

Gonnet, G. H., Cohen, M. A. & Benner, S. A. (1992). Exhaustive matching of the entire protein sequence database [see comments]. *Science* **256**, 1443-5.

Gribskov, M., McLachlan, A. D. & Eisenberg, D. (1987). Profile analysis: detection of distantly related proteins. *Proc Natl Acad Sci U S A* **84**, 4355-8.

Gribskov, M. & Veretnik, S. (1996). Identification of sequence pattern with profile analysis. *Methods Enzymol* **266**, 198-212.

Henikoff, S. & Henikoff, J. G. (1992). Amino acid substitution matrices from protein blocks. *Proc Natl Acad Sci U S A* **89**, 10915-9.

Karlin, S., Ost, F. & Blaisdell, B. E. (1989). Patterns in DNA and amino acid sequences and their statistical significance. In *Mathematical Methods for DNA Sequences* (ed. Waterman, M. S.), pp. 133-57. CRC P, Boca Raton.

Kosko, B. (1992). *Neural Networks and Fuzzy Systems*. Prentice-Hall, Englewood Cliffs.

Kyte, J. & Doolittle, R. F. (1982). A simple method for displaying the hydropathic character of a protein. *J Mol Biol* **157**, 105-32.

Lesser, G. J. & Rose, G. D. (1990). Hydrophobicity of amino acid subgroups in proteins. *Proteins* **8**, 6-13.

Nakata, K. (1995). Prediction of zinc finger DNA binding protein. *Comput Appl Biosci* **11**, 125-31.

Nevill-Manning, C. G., Wu, T. D. & Brutlag, D. L. (1998). Highly specific protein sequence motifs for genome analysis. *Proc Natl Acad Sci U S A* **95**, 5865-71.

Rice, D. W. & Eisenberg, D. (1997). A 3D-1D substitution matrix for protein fold recognition that includes predicted secondary structure of the sequence. *J Mol Biol* **267**, 1026-38.

Rose, G. D., Geselowitz, A. R., Lesser, G. J., Lee, R. H. & Zehfus, M. H. (1985). Hydrophobicity of amino acid residues in globular proteins. *Science* **229**, 834-8.

Rose, G. D. & Dworkin, J. E. (1989). The hydrophobicity profile. In *Prediction of Protein Structure and the Principles of Protein Conformation* (ed. Fasman, G. D.), pp. 625-33. Plenum P, New York.

Smith, R. F. & Smith, T. F. (1992). Pattern-induced multi-sequence alignment (PIMA) algorithm employing secondary structure-dependent gap penalties for use in comparative protein modeling. *Protein Eng* **5**, 35-41.

Sunyaev, S. R., Eisenhaber, F., Argos, P., Kuznetsov, E. N. & Tumanyan, V. G. (1998). Are knowledge-based potentials derived from protein structure sets discriminative with respect to amino acid types? *Proteins* **31**, 225-46.

Wu, C., Whitson, G., McLarty, J., Ermongkonchai, A. & Chang, T. C. (1992). Protein classification artificial neural system. *Protein Sci* **1**, 667-77.

Zadeh, L. A. (1965). Fuzzy sets. *Inform Control* **8**, 338-52.

CHAPTER 7

Design Issues – Data Encoding

As overviewed in the previous chapter (6.1), once a desired input/output mapping task is determined, the design of a complete system then involves the choices for a pre-processor and a post-processor, in addition to the neural network model itself. Pre-processing involves both feature representation (6.6) and input encoding. For molecular applications, the input sequence encoding method converts molecular sequences (character strings) into input vectors (numbers) of the neural networks. Likewise, an output encoding method is used in the post-processing step to convert the neural network output to desired forms.

Sequence encoding provides formalism for encoding protein sequences so that they can be processed using established methodologies in mathematics, statistics or machine learning. An ideal encoding scheme should extract maximal information from the sequence, and satisfy the basic coding assumption such that similar sequences are represented by *close* vectors. In this chapter we discuss the direct and indirect sequence encoding methods for extracting local and global features (7.1, 7.2), the construction of the input layer (7.3), feature analysis for input trimming (7.4), and output encoding methods (7.5).

7.1 Direct Input Sequence Encoding

Molecular applications involve the processing of individual residues, sequence windows of n-consecutive residues (termed *n-grams* or *k-tuples*), or the complete sequence string. Correspondingly, the encoding may be *local* (involving only single or neighboring residues in short sequence segment) or *global* (involving long-range relationship in full-length sequence or long sequence segment). The sequence encoding methods may be categorized as *direct encoding* or *indirect encoding*. Direct encoding converts each individual molecular residue to a vector (Table 7.1), whereas indirect encoding provides overall information measure of a complete sequence string (Table 7.2). Direct encoding preserves positional information, but can only deal with fixed-length sequence windows. On the other hand, indirect encoding disregards the ordering information, but can be used for sequences of either fixed or variable lengths.

In the direct encoding, each molecular residue can be represented by its identity or its features. The most commonly used method for direct encoding involves the use of *indicator vectors.* The indicator vector usually is a vector of binary numbers (0 or 1) that has only one unit turned on to indicate the *identity* of the corresponding residue. Here, a vector of four units with three zeros and a single one is needed for a nucleotide, so the four nucleotides may be represented as 1000 (A), 0100 (T), 0010 (G), and 0001 (C) (e.g., Brunak *et al.*, 1991). The spacer residue may be represented as 0000, without an additional unit (e.g., Larsen *et al.*, 1995). Likewise, a vector of 20 input units (nineteen zeros and a single one) is needed to represent an amino acid (e.g., Bohr *et al.*, 1988). Or a vector of 21 units may be used to include a unit for the spacer in regions between proteins (e.g., Qian & Sejnowski, 1988). These binary representations are dubbed as *BIN4*, *BIN20*, and *BIN21* (Table 7.1). The lengths of their input vectors are 4 x n, 20 x n, and 21 x n, respectively, where n is the total number of nucleotide or amino acid residues in the sequence window. The indicator vectors can be extended to other reduced alphabets (Table 6.2), and may be used for single residues, residue pairs or triples (two to three consecutive residues).

Table 7.1 *Direct sequence encoding methods*
for a single molecular residue and a 7-amino acid sequence window.

Encoding Method	Residue/ Window	Vector Size	Vector or Scalar Value*
Indicator Vector of AA Alphabet (BIN20)	Alanine	20	{1,0,0,0,0,0,0,0,0,0,0,0,0,0,0,0,0,0,0,0}
Ind. Vector of AA Alph. & Spacer (BIN21)	Alanine	21	{1,0,0,0,0,0,0,0,0,0,0,0,0,0,0,0,0,0,0,0,0}
Ind. Vector of Exchange Group (EG) Alph.	Alanine	6	{0, 0, 0, 1, 0, 0}
Ind. Vector of Hydrophobicity (HP) Alph.	Alanine	3	{0, 0, 1}
Fuzzy Vector of HP Alph.	Alanine	3	{0, 0.2, 0.8}
Feature Vector of K&D HP Scale	Alanine	1	{1.8}
PAM Substitution Vector	Alanine	20	Values varied
Evolutionary Vector	Alanine	20	Values varied
Ind. Vector of NA Alph. (Sparse) (BIN4)	Adenine	4	{1, 0, 0, 0}; others are 0100, 0010,0001
Ind. Vector of NA Alph. (Dense)	Adenine	2	{0, 0}; other are 01, 10, 11
Ind. Vector of NA Alph. (Ordinal)	Adenine	1	{1}; others are 2, 3, 4
Ind.Vector of AA Alph. (BIN20)	ANLAIDV	20 x 7	{1,19x 0;11x0,1,8x0; 9x0,…;17x0,1,0,0}
Ind. Vector of Exchange Group (EG) Alph.	ANLAIDV	6 x 7	{000100; 010000; 000010; …; 000010}
Ind. Vector of Hydrophobicity (HP) Alph.	ANLAIDV	3 x 7	{001; 100; 001; 001; 001; 100; 001}
Ind. Vector of HP Alph. Pairs	ANLAIDV	9 x 6	{000000100;001000000;….; 001000000}
Feature Vector of K&D HP Scale	ANLAIDV	1 x 7	{1,8; -3.5; 3.8; 1.8; 4.5; -3.5; 4.1}
PAM Substitution Vector	ANLAIDV	20 x 7	Values varied
Evolutionary Vector	ANLAIDV	20 x 7	Values varied

Ind. Vector = Indicator Vector; Alph. = Alphabet; AA = Amino Acid; EG = Exchange Group; HP = Hydrophobicity; NA = Nucleic Acid. The HP scale, alphabets, PAM substitution matrix, and evolution profile are as defined in Tables 6.1 and 6.2, and Figures 6.2 and 6.3, respectively.

*Raw data before scaling or normalization.

An alternative to the sparse encoding scheme of BIN4 is a dense representation that uses two units for four nucleotides (e.g., 00 for A, 01 for T, 10 for G, and 11 for C) (Demeler & Zhou, 1991). A comparative study, however, showed that the BIN4 coding was better, possibly due to its unitary coding matrix with identical Hamming distance among each vector. In Arrigo *et al.* (1991), the coding values were computed by dividing the ordinal number of each base (A=1; C=2; G=3; T=4) by the corresponding molecular weight, which resulted in the base values: A=0.0374; C=0.0822, G=0.1059 and T=0.1683.

In addition to the use of binary numbers to represent the identity of individual sequence residues, real numbers that characterize the residues can be used in the direct sequence encoding method. Each residue can be represented by a single feature, such as the hydrophobicity scale (Xin *et al.*, 1993), or by multiple properties that may or may not be orthogonal (Lohmann *et al.*, 1994). Each sequence position can also be represented by the residue frequency derived from multiple sequence alignments (i.e., sequence profile of a family) (Rost & Sander, 1993) or the substitution vector. The values of the vectors are usually normalized to a scale of 0 to 1, or -1 to 1 (Xin, 1993; Lohmann *et al.*, 1994).

7.2 Indirect Input Sequence Encoding

The direct sequence encoding methods preserve the order of residues along the sequence string and encode primarily local information. They are, however, not suitable when global sequence features or information content is more important to the application, or when variable length sequences are to be analyzed. This is evident in the intron/exon sequence discrimination (e.g., Uberbacher & Mural, 1991; Snyder & Stormo, 1993) and protein classification (e.g., Wu *et al.*, 1992; Ferran & Ferrara, 1992) problems.

A commonly used indirect sequence encoding method is the n-gram hashing method, which computes residue frequencies (Wu *et al.*, 1992). Residue frequency (composition) is one of the most effective global features for sequence discrimination. As shown in Figure 7.1, alternative alphabets of various n-gram terms can be used to maximize information extraction. The final size of the input vector is M^n, where M is the alphabet size (number of letters in the alphabet set), and n is the n-gram size (length of the sliding window along the sequence string) (Table 7.2). Note that because it is a hashing function, the n-gram method is more insertion/deletion invariant and can be used for sequences of variable lengths without pre-alignment of sequences.

A comparative study between the direct and indirect encoding methods was conducted by Farber *et al.* (1992) for coding region prediction. They observed that a dicodon frequency representation yielded significantly better results than an L x 64 string (indicator vector) representation for L-codon long fragments. (Note that the dicodon frequency is the 6-gram composition of the NA alphabet, with a vector size of 4096 or 4^6, and the BIN4 indicator representation of each 3-gram codon has a vector size of 64 or 4^3).

Other sequence features and information content can be extracted by various scoring mechanisms (e.g., Uberbacher *et al.*, 1996). Another example of indirect sequence encoding is the computation of hydrophobic moment (Chapter 6.4).

***Table 7.2** Indirect sequence encoding methods for a 7-amino acid sequence window.*

Encoding Method	Window	Vector Size	Vector or Scalar Value*
Hydrophobic Moment	ANLAIDV	1	Value varied
Hydropathy Index (of Central AA)	ANLAIDV	1	Value varied
Composition of AA Alph. (N-gram, A1)	ANLAIDV	20 (20^1)	{2;0;1;0;0;0;0;1;0;0;0;1;0;0;0;0;1;0;0}
Composition of HP Alph. (N-gram, H1)	ANLAIDV	3 (3^1)	{2; 0; 5}
Composition of HP Alph. Pair (N-gram, H2)	ANLAIDV	9 (3^2)	{0; 0; 2; 0; 0; 0; 2; 0; 2}

AA = Amino Acid; HP = Hydrophobicity; Alph. = Alphabet.

*Raw data before scaling or normalization.

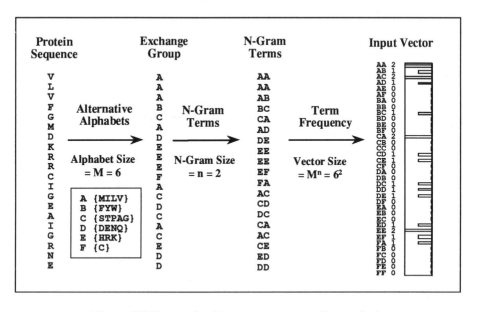

***Figure 7.1** N-gram hashing sequence encoding method.*

7.3 Construction of Input Layer

Deciding how to construct the input layer is application dependent. It is affected by many considerations. Should fixed-length sequence windows or variable-length sequences be used? Is there a dependence on positional information? Is it intended to search for signal or search for content? What is the importance of local information or global information?

The vector representation makes it easy to combine a large number of heterogeneous features without increasing the complexity of the system architecture. The input vectors representing various features can be concatenated to improve feature extraction and predictive accuracy. Figure 7.2 illustrates the use of a composite indicator vector for a 5-gram sequence window. Here, each amino acid residue is represented by 39 units, concatenated from binary indicator vectors of the different alphabet sets with 20, 6, 4, 3, 3, and 3 units. As discussed above, the indicator vector may also be real-valued, or derived from various alphabets.

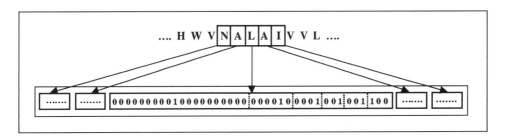

Figure 7.2 *A composite indicator vector for a five-gram sequence window.*

It is also possible to combine vectors from different (direct and indirect) encoding methods, such as the combination of indicator vectors and n-gram hash vectors, or to combine vectors of both local and global features. A more important consideration in combining the heterogeneous input vectors is the interdependency of these features and their effectiveness. Individual features can be examined separately along contiguous sequence windows to detect regions of unique features not obvious at the sequence level. This may be particularly important when multiple sequence alignments indicate extensive gaps within a structural motif. To determine whether two properties are orthogonal, one can also calculate the correlation coefficients (Lohmann *et al.*, 1994), as in

$$r = \frac{\displaystyle\sum_{i=1}^{20}(x_{i,1} - \overline{x}_1)(x_{i,2} - \overline{x}_2)}{\sqrt{\displaystyle\sum_{i=1}^{20}(x_{i,1} - \overline{x}_1)^2 \sum_{i=1}^{20}(x_{i,2} - \overline{x}_2)^2}}$$

where x_1 and x_2 are two different property scales. Feature extraction with redundant features is an important consideration for coding region recognition because the various coding measures have a great deal of redundancy. Some are sensing similar things, whereas many are derived from others. It was observed that often the direct measure provides better accuracy than those derived from it (Fickett, 1996).

The incorporation of different input features may also come in the form of different neural network designs. One example is cascading networks (Chapter 8), where output from one network becomes the input to the next network.

7.4 Input Trimming

An excessive number of input units not only results in poor generalization, but also slows network training. Reducing the number of input variables often leads to improved performance of a given data set, even though information is being discarded. Thus, one important role of pre-processing is to reduce the dimensionality of the input data.

One obvious way to reduce the number of input units is to use a feature detector that extracts salient features from the input data before presenting it to the neural networks. Several methods have been devised for molecular studies. A decision tree algorithm was used for feature selection from large candidate pools (Cherkauer & Shavlik, 1993). A *singular value decomposition* (SVD) procedure (Wu *et al.*, 1995) and the statistical method of *principal component analysis* (PCA) (Ferran & Pflugfelder, 1993) were used for a dimensional reduction of the input data. The SVD and PCA methods are closely related, both of which extract orthogonal vectors in a lower dimensional space without losing essential intrinsic information. Alternatively, the input units may be trimmed during training using *network pruning* procedures. This was done in Horton and Kanehisa (1992), where the least important input units were removed, as reflected by their low absolute weights (i.e., with the idea of *magnitude equals saliency*).

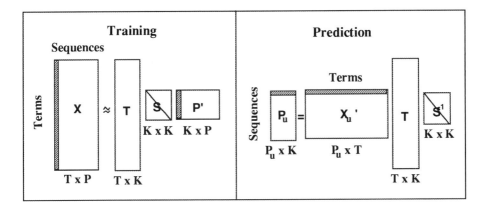

Figure 7.3 *Singular value decomposition.*

Singular Value Decomposition. SVD is used to reduce the size of n-gram vectors and to extract semantics from the n-gram patterns. The method was adopted from the *Latent Semantic Indexing* (LSI) analysis used in the field of information retrieval and information filtering. The LSI approach is to take advantage of implicit high-order structure in the association of terms with documents in order to improve the detection of relevant documents, which may or may not contain actual query terms. In SVD (Figure 7.3), the term matrix (i.e., term-by-sequence matrix) is decomposed into a set of *k* orthogonal factors from which the original matrix can be approximated by linear combination. The reduced model can be shown by

$$\mathbf{X} \cong \mathbf{Y} = \mathbf{TSP'} \qquad (7.1)$$

where **X** is the original term matrix of rank *m* (*m* < min (*t*,*p*)), **Y** is the approximation of **X** with rank *k* (*k* < *m*), **T** and **P** are the matrices of left and right singular (*s*) vectors corresponding to *k*-largest *s* values, and **S** is the diagonal matrix of *k*-largest *s* values. Note that (a) both **T** and **P** have orthonormal columns, and (b) if **X** is used to represent the original term matrix for training sequences, then **P** becomes the reduced matrix for the training sequences. The representation of unknown sequences is computed by *folding* them into the k-dimensional factor space of the training sequences. The folding technique, which amounts to placing sequences at the centroid of their corresponding term points, can be expressed by

$$\mathbf{P_u} = \mathbf{X_u'TS}^{-1} \qquad (7.2)$$

where $\mathbf{P_u}$ and $\mathbf{X_u}$ are the reduced and original term matrices of unknown sequences, **T** is the matrix of left s-vectors computed from equation 7.1 during the training phase, and $\mathbf{S}^{-1}$

is the inverse of **S**, which reflects scaling by reciprocals of corresponding *s* values. It has been shown that the SVD method also satisfies the basic coding assumption (Wu *et al.*, 1995).

7.5 Output Encoding

The choice of output encoding method is more straightforward, depending on the neural network mapping desired. The discrimination problem usually involves a yes/no answer that can be encoded using a single output unit (*dense encoding*) (e.g., Stormo *et al.*, 1982) or two output units (*sparse encoding*) (e.g., Fariselli & Casadio, 1996). The number of output units in the classifier is usually the same as the number of classes, encoded as indicator vectors, such as the number of secondary structure states (e.g., Qian & Sejnowski, 1988) or protein families (e.g., Wu *et al.*, 1992). Some unsupervised algorithms, such as the adaptive resonance theory, self-organize the network and configure the number of output units (i.e., categories) automatically (LeBlanc *et al.*, 1994). The value of the output units can be used qualitatively (all-or-none by comparing to a threshold value), or as a quantitative measure of confidence level (Chandonia & Karplus, 1995) or activity level (e.g., Nair *et al.*, 1995).

7.6 References

Arrigo, P., Giuliano, F., Scalia, F., Rapallo, A. & Damiani, G. (1991). Identification of a new motif on nucleic acid sequence data using Kohonen's self organizing map. *Comput Appl Biosci* **7**, 353-7.

Bohr, H., Bohr, J., Brunak, S., Cotterill, R. M., Lautrup, B., Norskov, L., Olsen, O. H. & Petersen, S. B. (1988). Protein secondary structure and homology by neural networks. The alpha-helices in rhodopsin. *FEBS Lett* **241**, 223-8.

Brunak, S., Engelbrecht, J. & Knudsen, S. (1991). Prediction of human mRNA donor and acceptor sites from the DNA sequence. *J Mol Biol* **200**, 49-66.

Chandonia, J. M. & Karplus, M. (1995). Neural networks for secondary structure and structural class predictions. *Protein Sci* **4**, 275-85.

Cherkauer, K. J. & Shavlik, J. W. (1993). Protein structure prediction: selecting salient features from large candidate pools. *Ismb* **1**, 74-82.

Demeler, B. & Zhou, G. W. (1991). Neural network optimization for *E. coli* promoter prediction. *Nucleic Acids Res* **19**, 1593-9.

Farber, R., Lapedes, A. & Sirotkin, K. (1992). Determination of eukaryotic protein coding regions using neural networks and information theory. *J Mol Biol* **226**, 471-9.

Fariselli, P. & Casadio, R. (1996). HTP: a neural network-based method for predicting the topology of helical transmembrane domains in proteins. *Comput Appl Biosci* **12**, 41-8.

Ferran, E. A. & Ferrara, P. (1992). Clustering proteins into families using artificial neural networks. *Comput Appl Biosci* **8**, 39-44.

Ferran, E. A. & Pflugfelder, B. (1993). A hybrid method to cluster protein sequences based on statistics and artificial neural networks. *Comput Appl Biosci* **9,** 671-80.

Fickett, J. W. (1996). The gene identification problem: An overview for developers. *Comput Chem* **20,** 103-18.

Horton, P. B. & Kanehisa, M. (1992). An assessment of neural network and statistical approaches for prediction of E. coli promoter sites. *Nucleic Acids Res* **20,** 4331-8.

Larsen, N. I., Engelbrecht, J. & Brunak, S. (1995). Analysis of eukaryotic promoter sequences reveals a systematically occurring CT-signal. *Nucleic Acids Res* **23,** 1223-30.

LeBlanc, C., Katholi, C. R., Unnasch, T. R. & Hruska, S. I. (1994). DNA sequence analysis using hierarchical ART-based Classification Networks. *Ismb* **2,** 253-60.

Lohmann, R., Schneider, G., Behrens, D. & Wrede, P. (1994). A neural network model for the prediction of membrane-spanning amino acid sequences. *Protein Sci* **3,** 1597-601.

Nair, T. M., Tambe, S. S. & Kulkarni, B. D. (1995). Analysis of transcription control signals using artificial neural networks. *Comput Appl Biosci* **11,** 293-300.

Qian, N. & Sejnowski, T. J. (1988). Predicting the secondary structure of globular proteins using neural network models. *J Mol Biol* **202,** 865-84.

Rost, B. & Sander, C. (1993). Prediction of protein secondary structure at better than 70% accuracy. *J Mol Biol* **232,** 584-99.

Snyder, E. E. & Stormo, G. D. (1993). Identification of coding regions in genomic DNA sequences: an application of dynamic programming and neural networks. *Nucl Acids Res* **21,** 607-13.

Stormo, G. D., Schneider, T. D., Gold, L. & Ehrenfeucht, A. (1982). Use of the 'Perceptron' algorithm to distinguish translational initiation sites in *E. coli*. *Nucleic Acids Res* **10,** 2997-3011.

Uberbacher, E. C. & Mural, R. J. (1991). Locating protein-coding regions in human DNA sequences by a multiple sensor-neural network approach. *Proc Natl Acad Sci U S A* **88,** 11261-5.

Uberbacher, E. C., Xu, Y. & Mural, R. J. (1996). Discovering and understanding genes in human DNA sequence using GRAIL. *Methods Enzymol* **266,** 259-81.

Wu, C., Whitson, G., McLarty, J., Ermongkonchai, A. & Chang, T. C. (1992). Protein classification artificial neural system. *Protein Sci* **1,** 667-77.

Wu, C. H., Berry, M., Shivakumar, S. & McLarty, J. (1995). Neural networks for full-scale protein sequence classification: sequence encoding with singular value decomposition. *Machine Learning* **21,** 177-93.

Xin, Y., Carmeli, T. T., Liebman, M. N. & Wilcox, G. L. (1993). Use of the backpropagation neural network algorithm for prediction of protein folding patterns. In *Proceedings of the Second International Conference on Bioinformatics, Supercomputing, and Complex Genome Analysis* (ed. Lim, H. A., Fickett, J. W., Cantor, C. R., and Robbins, R.J.), pp. 359-75. . World Scientific, New Jersey

CHAPTER 8

Design Issues – Neural Networks

There is a fundamental difference between the design of a neural network and that of its classical information-processing counterpart. In the latter case, one usually proceeds by first formulating a mathematical model of environment observations, validating the models with real data, and then building the design on the basis of the model. In contrast, the design of a neural network is based directly on real data. Thus, the neural network not only provides an implicit model of the environment in which it is embedded, but also performs the information-processing function of interest. In a neural network of specified architecture, knowledge presentation of the surrounding environment is defined by the values taken on by the free parameters (i.e., *weights*) of the network. The form of this knowledge representation constitutes the very design of the neural network, and therefore holds the key to its performance.

An important way to improve network performance is through the use of *prior knowledge*, which refers to information that one has about the desired form of the solution and which is additional to the information provided by the training data. Prior knowledge can be incorporated into the pre-processing and post-processing stages (Chapter 7), or into the network structure itself.

Examined below are several neural network design considerations, including the architecture (8.1), learning algorithm (8.2), network parameters (8.3), training and test data (8.4), and evaluation mechanism (8.5).

8.1 Network Architecture

Some network architecture choices include the number of layers, the number of processing units in each layer, the specific interconnections between the units, and the number of neural networks. One may use a perceptron with a two-layer topology if the input data is *linearly separable*. The major advantages of perceptrons are their convergence properties, their simplicity and rule extraction. The perceptron has few interconnections; thus, it requires less training time to adjust the weights, and needs fewer training patterns to achieve good generalization. The weight matrix of a perceptron can be represented by a *Hinton graph* and used to decipher biological information or derive consensus sequence patterns (e.g., Qian & Sejnowski, 1988). No gain over a perceptron would be realized by a multilayer perceptron network, when there is limited correlation information between input data (Stolorz *et al.*, 1992), or when there is not sufficient input to saturate the more complex network (Snyder & Stormo, 1995). On the other hand, if nonlinear input-output mapping is desired, then a multilayer perceptron with at least one hidden layer is required. The ability of hidden neurons to extract higher-order statistics is particularly valuable when the size of the input layer is large. Unfortunately, no convergence guarantee exists for multilayer perceptrons, which have nonlinear outputs.

The multilayer perceptron can approximate any arbitrary mapping only when the network is sufficiently large. But if the network is too big, the learning problem becomes unconstrained and results in poor generalization. Instead, *minimal architecture* (with fewer neurons and interconnections) should be used to avoid data *overfitting*. This design objective can be achieved in different ways. In *network growing*, one starts with a small network and allows it to grow gradually by adding new units whenever design specification is not met. The *cascade-correlation* learning architecture (Fahlman & Lebiere, 1990) is an example of the network growing approach. In *network pruning*, one starts with a large network with an adequate performance for the problem at hand, and then prunes it by eliminating excessive weights and/or neurons (surveyed in Reed, 1993). One approach to the network pruning is through the use of *weight decay* (Hinton, 1987), in which each weight decays toward zero at a rate proportional to its magnitude. Another pruning approach is to remove weights on the basis of their saliency, as exemplified by the so-called *optimal brain damage* (Le Cun *et al.*, 1990b) and the *optimal brain surgeon* (Hassibi *et al.*, 1993). Algorithms have also been developed to determine the network topology automatically, such as the *dynamic node architecture learning* (Bartlett, 1994) based on information theory.

The processing units between layers can be fully or partially connected. It is generally agreed that it is inadvisable for a multilayer perceptron to be fully connected. Instead, prior information should be built into the design of a neural network, thereby simplifying the network design. This results in a neural network with a specialized (restricted) structure and is highly desirable for several reasons. In particular, the restricted network usually has fewer free parameters available for adjustment than a fully connected network. Consequently, the specialized network requires a smaller data set for training, learns faster and often generalizes better. Such a design is well illustrated by the optical character recognition problem (Le Cun *et al.*, 1990a). Here, prior information is built into the neural network design by combining two techniques: (1) restricting the network architecture through the use of *local connections*, and (2) constraining the choice of synaptic weights by the use of *weight sharing* where several weights are controlled by a single parameter. Though most networks used in the molecular studies are fully connected, local connections are used in Snyder and Stormo (1995).

A further approach to designing the network topology is to construct a complex network from several single network modules. In a *network committee*, several networks are trained together to form a committee (Perrone, 1994). A *committee* or *jury decision*, given by a weighted combination of the predictions of the members, can yield better performance than the performance of the best single network used in isolation. The jury decision can be used to compromise between over-prediction and under-prediction made by neural networks of different architectures or trained by different strategies. For example, a jury (with a simple arithmetic average and winner-take-all) over four neural networks (with balanced or unbalanced training strategies) was used to address the problem of uneven distribution of training classes (Rost, 1996).

In a *mixture-of-experts* model (Jacobs *et al.*, 1991), different expert networks were assigned to tackle sub-tasks of training cases, and an extra gating network was used to decide which of the experts should determine the output. The model discovered a suitable decomposition of the input space as part of the learning process. Later the model was further extended (Jordan & Jacobs, 1994) into a hierarchical system with a tree structure. In molecular applications, cascaded networks--where outputs of some networks become the inputs of others--were used to improve performance (Rost & Sander, 1994). Multiple neural network modules may run in parallel in order to scale up the system (Wu *et al.*, 1995). More than one network can also be used to extract different (e.g., local vs. global) features (Mahadevan & Ghosh, 1994).

8.2 Network Learning Algorithm

The choice of a *supervised* or *unsupervised* learning algorithm mainly depends on the availability of an external teacher and the tasks to be solved. The *supervised* training is accomplished by presenting a sequence of training vectors, each with an associated *target* output vector. In conceptual terms, we may think of the teacher as having knowledge of the environment that is represented by a set of input-output examples. Knowledge of the environment available to the teacher is transferred to the neural networks through iterative adjustments to minimize the error signal according to a learning algorithm. The sequence discrimination problems normally involve supervised networks because of the existence of examples and counter-examples (e.g., coding and non-coding sequences, promoter and non-promoter sequences). Similarly, the protein structure prediction and classification problems with known categories also use supervised networks (e.g., α helix, β sheet and random coil; or protein superfamilies).

In *unsupervised* or *self-organized* learning there is no external teacher to oversee the learning process. The learning normally is driven by a *similarity* measure without specifying target vectors. The self-organizing net modifies the weights so that the most similar vectors are assigned to the same output (cluster) unit that is represented by an *examplar vector*. For genome informatics applications, unsupervised learning can be used for feature extraction (e.g., Arrigo *et al.*, 1991; Giuliano *et al.*, 1993), clustering (e.g., Ferran *et al.*, 1994), or classification (e.g., LeBlanc *et al.*, 1994).

Supervised and unsupervised learning may coexist in a given architecture. The *counter-propagation* network (Hecht-Nielsen, 1987) is an example in which layers from supervised and unsupervised learning paradigms are combined to construct a new type of network. A modified counter-propagation network, which employed a supervised *learning vector quantizer* algorithm to perform nearest-neighbor classification, was used for molecular sequence classification (Wu *et al.*, 1997). Some applications may benefit from using combined output from multiple neural networks trained with different learning algorithms. For example, the combination of back-propagation and counter-propagation networks resulted in better classification accuracy than that of either method alone (Wu & Shivakumar, 1994).

Hybrid systems may also be developed to combine a neural network approach with other designs, such as knowledge-based systems, machine learning, genetic algorithms, fuzzy logic and statistical methods (Kandel & Langholz, 1992; Fu, 1994). Several hybrid systems have been used for molecular sequence analysis. A *knowledge-based* approach was used to apply inference rules about the current biological problem to configure neural networks (Shavlik *et al.*, 1992; Maclin & Shavlik, 1993). A neural network was used as a *fitness function* for genetic algorithms (Schneider *et al.*, 1994). Conversely, a *genetic algorithm* was used to optimize a neural network topology (Vivarelli *et al.*, 1995), and as an alternative neural network learning algorithm (Lohmann *et al.*, 1994).

8.3 Network Parameters

To optimize the neural network design, important choices must be made for the selection of numerous parameters. Many of these are internal parameters that need to be tuned with the help of experimental results and experience with the specific application under study. The following discussion focuses on back-propagation design choices for the learning rate, momentum term, activation function, error function, initial weights, and termination condition.

The selection of a *learning rate* (i.e., step size in the minimization process) is important in finding the true global minimum of the error distance. The convergence speed of back-propagation is directly related to the learning rate parameter. Too small a learning rate will make the training progress slow, whereas too large a learning rate may simply produce oscillations between relatively poor solutions. In general, large steps are desirable when the search point is far away from a minimum, with a decreasing step size as the search approaches a minimum. A sample of various approaches for selecting the proper learning rate is given in Hassoun (1995). A recent study (Salomon & van Hemmen, 1996) presented the use of a new genetic algorithm, *dynamic self-adaptation*, to adjust the learning rate to the landscape generated by the error function. The procedure improved the convergence rate by several orders of magnitude as compared to standard back-propagation.

A *momentum term* can be helpful in speeding convergence and avoiding *local minima* (Rumelhart *et al.*, 1986). In back-propagation with momentum, the weight change is in a direction that is a combination of the current gradient and the previous gradient. This is a modification of gradient descent in which advantages arise chiefly when some training data are very different from the majority of the data (and possibly even incorrect). Momentum allows the net to make reasonably large weight adjustments as long as the corrections are in the same general direction for several patterns, while using a smaller learning rate to prevent a large response to the error from any one training pattern. It also reduces the likelihood that the net will find weights that represent a local minimum. The momentum term is normally chosen between 0 and 1. Adaptive momentum rates also may be employed. Fahlman (1988) proposed a heuristic variation of back-propagation, called *QuickProp*, which employed a dynamic momentum rate.

An *activation function* for a back-propagation net should satisfy several important characteristics. It should be *continuous* and *differentiable*. For computational efficiency, it should have a derivative that is easy to compute. Common activation functions are discussed in Chapter 3.2.

A commonly used *error function* is the *root-mean-square error*, which is the square root of the sum-of-square errors calculated from all patterns across the entire training file. Other error functions (*cost functions*) may also be defined (Van Ooyen & Nienhuis, 1992; Rumelhart *et al.*, 1995), depending on the particular application.

Owing to its gradient-descent nature, back-propagation is very sensitive to initial conditions. The choice of *initial weights* will influence whether the net reaches a global (or only a local) minimum of the error and, if so, how quickly it converges. In practice, the weights are usually initialized to small zero-mean random values between -0.5 and 0.5 (or between -1 and 1 or some other suitable interval).

The back-propagation algorithm cannot, in general, be shown to converge, nor are there well-defined *stopping criteria*. A few reasonable criteria exist which may be used to terminate the weight adjustment. Since the usual motivation of applying a back-propagation net is to achieve a balance between correct response to training patterns and good responses to new input patterns, it is not necessarily advantageous to continue training until the total squared error actually reaches a minimum. The training is usually stopped by a user-determined threshold value (called *tolerance*) for the error function, or a fixed upper limit on the number of training iterations (called *epochs*). The termination can also be based on performance of a validation data set used to monitor generalization performance during learning and to terminate learning when there is no more improvement.

was modified to reduce the back-propagation training time by several folds with similar predictive accuracy (Wu & Shivakumar, 1994).

The neural network training itself may give clues to the integrity of the data. By carefully monitoring the learning process, one often finds that noisy or wrong data is either very slow to train or completely untrainable. When available, larger data sets provide better statistics and allow finer-grain classification either through the sub-grouping of data or a multi-step classification. Better performance may also be resulted. In a recent study involving a 32-fold cross-validated test for secondary structure and structural class prediction, Chandonia & Karplus (1996) showed that, for the larger database, better results were obtained with more units in the hidden layer.

8.4.3 Benchmarking Data Set

Comparative performance evaluation of various computational methods requires the use of standard data sets as benchmarks. Different benchmarks have been developed for various genome informatics applications. For the gene recognition methods, commonly used is the Burset and Guigo (1996) set of 570 vertebrate multi-exon gene sequences. The data set includes only short sequences for encoding a single complete gene with simple structure, high coding density, and no sequencing errors. For human genome annotation, a cross-validated standard test set of 304 human genes (Kulp *et al.*, 1996) is available.

For protein secondary structure and structural class prediction, the set of 126 non-homologous globular protein chains compiled by Rost & Sander (1993) from the Brookhaven Protein Data Bank (PDB) (Bernstein *et al.*, 1977) is often used as a benchmark. To ensure a reasonable assessment of the method's generalization ability, the representative set consisted of only structures without significant homology (i.e., < 25% pairwise similarity for chains with more than 80 residues) and of high resolution (i.e., at least 2.5A). Eight secondary classes were derived based on the known tertiary structures using the DSSP program (Kabsch & Sander, 1983), and then reduced into three categories of secondary structures. The three categories were α helix to represent H (α-helix), G (3_{10}-helix), and I (π-helix); β sheet to represent E (extended strand), and B (β-bridge); and loop (L) to represent T (3,4,5 turn), S (bend), and _ (rest, coil). The complete database thus contains a total of 24,395 residues with a composition of 32% α helix, 21% β sheet and 47% loop.

Later, a larger set of 318 non-homologous chains representative of high-resolution structures became available (Sali & Overington, 1994). All X-ray determined structures had a resolution of 2.3A or better (although some NMR-determined structures were also used), and no protein chains contained more than 30% sequence identity. All residues assigned by the DSSP program that were neither α helix (H) nor extended β sheet (E) were considered to be in the coil category, which included 3_{10}-helix (G). The complete data set has 56,966 residues, with 30% α helix, 21% β sheet and 49% coil.

The commonly used benchmark for protein fold recognition is the test set suggested by Fischer et al. (1996), which comprises a set of 68 pairs of proteins with very low sequence similarity, but very highly similar folds.

8.5 Evaluation Mechanism

Cross-validation is the standard method for evaluating generalization performance with training and prediction sets. In *k*-fold cross-validation, the cases are randomly divided into *k* mutually exclusive test partitions of approximately equal size. The cases not found in each test partition are independently used for training, and the resulting network is tested on the corresponding test partition. The *leave-one-out* type of cross-validation is suitable for problems where small sample sizes are available. Here, a network of sample size *n* is trained using *n-1* cases and tested on the single remaining case, then repeated for each case in turn.

The next issue of evaluation mechanism is to define a measure for the quality of a particular prediction. For sequence discrimination problem, which involves a yes or no answer (e.g., binding or non-binding, member or non-member), the performance can be measured by sensitivity, specificity, positive predictive value, negative predictive value, accuracy, and correlation coefficient, as shown in

$$\text{Sensitivity} = \frac{TP}{TP + FN}$$

$$\text{Specificity} = \frac{TN}{TN + FP}$$

$$\text{Positive Predictive Value} = \frac{TP}{TP + FP}$$

$$\text{Negative Predictive Value} = \frac{TN}{TN + FN}$$

$$\text{Accuracy} = \frac{TP + TN}{Total}$$

$$\text{Correlation Coefficient} = \frac{(TP \times TN - FP \times FN)}{\sqrt{(TP + FP) \times (FP + TN) \times (TN + FN) \times (FN + TP)}}$$

where *TP* is true positive, *TN* is true negative, *FP* is false positive, and *FN* is false negative. Here, sensitivity is the proportion of all true positive patterns that are correctly identified, and specificity is the percentage of all true negative patterns that are correctly identified. They can be considered as a measure of how well false negatives and false positives are eliminated, respectively. The positive predictive value is the probability that a predicted true pattern is indeed a true pattern, whereas the negative predictive value is the probability that a predicted negative pattern is indeed a negative pattern. The accuracy is the percentage of all correct predictions. The correlation coefficient was introduced by Matthews (1975). Its value of 1 and -1 correspond to a perfect and a completely wrong prediction, respectively.

In protein secondary structure prediction, where a three-category (α, β, and coil or loop) prediction is made, the accuracy can be measured by a 3 x 3 accuracy table, as in Rost and Sander (1993).

Table 8.1 *A 3x3 accuracy matrix for evaluating protein secondary structure prediction.*

A	α	β	L	Observed
α	$A_{\alpha\alpha}(TP_\alpha)$	$A_{\alpha\beta}(FN_\alpha)$	$A_{\alpha L}(FN_\alpha)$	b_α
β	$A_{\beta\alpha}(FP_\alpha)$	$A_{\beta\beta}(TN_\alpha)$	$A_{\beta L}(TN_\alpha)$	b_β
L	$A_{L\alpha}(FP_\alpha)$	$A_{L\beta}(TN_\alpha)$	$A_{LL}(TN_\alpha)$	b_L
Predicted	a_α	a_β	a_L	N

$$a_i = \sum_{j=1}^{3} A_{ji}, \quad b_i = \sum_{j=1}^{3} A_{ij} \quad for\ i = \alpha, \beta, L$$

$$N = \sum_{j=1}^{3} a_j = \sum_{j=1}^{3} b_j$$

$$Q_i = Q_i^{\%obs} = \frac{A_{ii}}{b_i} \times 100, \quad Q_i^{\%pred} = \frac{A_{ii}}{a_i} \times 100 \quad for\ i = \alpha, \beta, L$$

$$Q_3 = \frac{\sum_{i=1}^{3} A_{ii}}{N} \times 100$$

where A_{ij} is the number of residues predicted to be in structure i and observed in structure j; a_i and b_i are predicted and observed frequency (number of residues) in a given structure; N is the total number of residues in the data set; Q_i is the percentage of residues correctly predicted to observed in a given structure; $Q_i^{\%pred}$ is the percentage of residues correctly predicted from all residues predicted in a given structure; and Q_3 is the three-state overall per-residue accuracy, measured as the percentage of all correctly predicted residues.

The 3 x 3 accuracy table can also be reduced to a 2 x 2 table for residues in a given class and in all other classes, and used to calculate the correlation coefficient.

$$TP_i = A_{ii}, \quad FN_i = \sum_{j \neq i}^{3} A_{ij}, \quad FP_i = \sum_{j \neq i}^{3} A_{ji}, \quad TN_i = \sum_{j \neq 1}^{3}\sum_{k \neq 1}^{3} A_{jk} \qquad for\, i = \alpha, \beta, L$$

$$C_i = \frac{(TP_i \times TN_i - FP_i \times FN_i)}{\sqrt{(TP_i + FP_i) \times (FP_i + TN_i) \times (TN_i + FN_i) \times (FN_i + TP_i)}}$$

Another measure for accuracy is an entropy-related information, I, as defined below. It merges the different percentages to a single number with all elements of the accuracy matrix (Table 8.1) contributing equally. The advantage of such entropy is that over- and under-predictions equally decrease the value of I. The normalized information value I equals zero in a randomly distributed population (where $A_{ij} = N/9$), whereas $I = 1$ in a perfect match between predicted and observed populations ($A_{ii} = b_i$ and $A_{ij} = 0$ for $i \mathrel{!}= j$).

$$I = 1 - \frac{\sum_{i=1}^{3} a_i \times \ln a_i - \sum_{ij=1}^{3} A_{ij} \times \ln A_{ij}}{N \times \ln N - \sum_{i=1}^{3} b_i \times \ln b_i}$$

8.6 References

Arrigo, P., Giuliano, F., Scalia, F., Rapallo, A. & Damiani, G. (1991). Identification of a new motif on nucleic acid sequence data using Kohonen's self organizing map. *Comput Appl Biosci* **7**, 353-7.

Bartlett, E. B. (1994). Dynamic node architecture learning: An information theoretic approach. *NEUNET* **7**, 129-40.

Baum, E. B. & Haussler, D. (1989). What size net gives valid generalization? *Neural Comput* **1**, 151-60.

Bernstein, F. C., Koetzle, T. F., Williams, G. J., Meyer, E. E. Jr., Brice, M. D., Rodgers, J. R., Kennard, O., Shimanouchi, T. & Tasumi, M. (1977). The Protein Data Bank: a computer-based archival file for macromolecular structures. *J Mol Biol* **112,** 535-42.

Burset, M. & Guigo, R. (1996). Evaluation of gene structure prediction programs. *Genomics* **34,** 353-67.

Cachin, C. (1994). Pedagogical pattern selection strategies. *NEUNET* **7,** 175-81.

Chandonia, J. M. & Karplus, M. (1996). The importance of larger data sets for protein secondary structure prediction with neural networks. *Protein Sci* **5,** 768-74.

DeRouin, E., Brown, J., Beck, H., Fausett, L. & Schneider, M. (1991). Neural network training on unequally represented classes. In: *Intelligent Engineering Systems Through Artificial Neural Networks* (ed. Dagli, C. H., Kumara, S. R. T. & Shin, Y. C.), pp. 135-41. ASME P, New York.

Fahlman, S. E. (1988). Fast learning variations on back-propagation: An empirical study. In *Proceedings of the 1988 Connectionist Models Summer School* (ed. Hinton, G. E., Sejnowski, T. J. & Touretzky, D. S.), pp. 38-51. Morgan Kaufmann, San Mateo, CA.

Fahlman, S. & Lebiere, C. (1990). The cascade-correlation learning architecture. *Adv Neural Inf Process Syst* **2,** 524-32.

Ferran, E. A., Pflugfelder, B. & Ferrara, P. (1994). Self-organized neural maps of human protein sequences. *Protein Sci* **3,** 507-21.

Fischer, D., Elofsson, A., Rice, D. & Eisenberg, D. (1996). Assessing the performance of fold recognition methods by means of a comprehensive benchmark. *Pac Symp Biocomput*, 300-18.

Fu, L. (1994). *Neural Networks in Computer Intelligence.* McGraw-Hill, New York.

Giuliano, F., Arrigo, P., Scalia, F., Cardo, P. P. & Damiani, G. (1993). Potentially functional regions of nucleic acids recognized by a Kohonen's self-organizing map. *Comput Appl Biosci* **9,** 687-93.

Hassibi, B., Stork, D. G. & Wolff, G. J. (1993). Optimal brain surgeon and general network pruning. *Proc IEEE International Conf Neural Networks* **1,** 293-9.

Hassoun, M. H. (1995). *Fundamentals of Artificial Neural Networks.* MIT P, Cambridge.

Hecht-Nielsen, R. (1987). Counterpropagation networks. *Applied Optics* **26,** 4979-84.

Hinton, G. E. (1987). Learning translation invariant recognition in a massively parallel network. In PARLE: Parallel Architecture and Languages (ed. Goos, G. & Hartmanis, J.), pp. 1-13. Springer-Verlag, Berlin.

Hobohm, U., Scharf, M., Schneider, R. & Sander, C. (1992). Selection of representative protein data sets. *Protein Sci* **1,** 409-17.

Jacobs, R. A., Jordan, M. I., Nowlan, S. J. & Hinton, G. E. (1991). Adaptive mixtures of local experts. *Neural Comput* **3,** 79-87.

Jordan, M. I. & Jacobs, R. A. (1994). Hierarchical mixtures of experts and the EM algorithm. *Neural Comput* **6,** 181-214.

Kabsch, W. & Sander, C. (1983). Dictionary of protein secondary structure: pattern recognition of hydrogen-bonded and geometrical features. *Biopolymers* **22,** 2577-637.

Kandel, A. & Langholz, G. (Eds.). (1992). *Hybrid Architectures for Intelligent Systems.* CRC P, Boca Raton.

Korning, P. G., Hebsgaard, S. M., Rouze, P. & Brunak, S. (1996). Cleaning the GenBank *Arabidopsis thaliana* data set. *Nucleic Acids Res* **24,** 316-20.

Kulp, D., Haussler, D., Reese, M. G. & Eeckman, F. H. (1996). A generalized hidden Markov model for the recognition of human genes in DNA. *Ismb* **4,** 134-42.

Lebeda, F. J., Umland, T. C., Sax, M. & Olson, M. A. (1998). Accuracy of secondary structure and solvent accessibility predictions for a clostridial neurotoxin C-fragment. *J Protein Chem* **17,** 311-8.

LeBlanc, C., Katholi, C. R., Unnasch, T. R. & Hruska, S. I. (1994). DNA sequence analysis using hierarchical ART-based Classification Networks. *Ismb* **2,** 253-60.

Le Cun, Y., Boser, B., Denker, J., Henderson, R. E., Howard, W., *et al.* (1990a). Handwritten digit recognition with a back-propagation network. *Adv Neural Inf Process Syst* **2,** 396-404.

Le Cun, Y., Denker, J. & Solla, S. (1990b). Optimal brain damage. *Adv Neural Inf Process Syst* **2,** 598-605.

Lohmann, R., Schneider, G., Behrens, D. & Wrede, P. (1994). A neural network model for the prediction of membrane-spanning amino acid sequences. *Protein Sci* **3,** 1597-601.

Maclin, R., and Shavlik J.W. (1993). Using knowledge-based neural network to improve algorithms: Refining the Chou-Fasman algorithm for protein folding. *Machine Learning* **11,** 195-215.

Mahadevan, I. & Ghosh, I. (1994). Analysis of *E.coli* promoter structures using neural networks. *Nucleic Acids Res* **22,** 2158-65.

Matthews, B. W. (1975). Comparison of the predicted and observed secondary structure of T4 phage lysozyme. *Biochim Biophys Acta* **405,** 442-51.

Morgan, N. & Bourland, H. (1990). Generalization and parameter estimation in feedforward nets: Some experiments. *Adv Neural Inf Process Syst* **2,** 630-7.

Perrone, M. P. (1994). General averaging results for convex optimization. In: *Proceedings 1993 Connectionist Models Summer School* (ed. Mozer, M. C., *et al.*), pp. 364-71. Lawrence Erlbaum, Hillsdale, NJ.

Qian, N. & Sejnowski, T. J. (1988). Predicting the secondary structure of globular proteins using neural network models. *J Mol Biol* **202,** 865-84.

Reed, R. (1993). Pruning algorithms - A survey. *IEEE Trans. Neural Networks* **4,** 740-7.

Rost, B. (1996). PHD: predicting one-dimensional protein structure by profile-based neural networks. *Methods Enzymol* 266, 525-39.

Rost, B. & Sander, C. (1993). Prediction of protein secondary structure at better than 70% accuracy. *J Mol Biol* **232,** 584-99.

Rost, B. & Sander, C. (1994). Combining evolutionary information and neural networks to predict protein secondary structure. *Proteins* **19,** 55-72.

Rumelhart, D. E., Hinton, G. E. & Williams, R. J. (1986). Learning representations by backpropagating errors. *Nature* **323,** 533-6.

Rumelhart, D. E., Durbin, R., Golden, R. & Chauvin, Y. (1995). Backpropagation: The basic theory. In *Backpropagation: Theory, Architectures and Applications*, (ed. Chauvin, Y. & Rumelhart, D. E.), pp. 1-34. Lawrence Erlbaum Associates, Hillsdale, NJ.

Sali, A. & Overington, J. P. (1994). Derivation of rules for comparative protein modeling from a database of protein structure alignments. *Protein Sci* **3,** 1582-96.

Salomon, R. & Van Hemmen, J. L. (1996). Accelerating backpropagation through dynamic self-adaptation. *NEUNET* **9,** 589-601.

Schneider, G., Schuchhardt, J., and Wrede, P. (1994). Artificial neural networks and simulated molecular evolution are potential tools for sequence-oriented protein design. *Comput Appl Biosci* **10,** 635-45.

Shavlik, J. W., Towell, G. G. & Noordewier, M. O. (1992). Using knowledge-based neural networks to refine existing biological theories. In *The Second International Conference on Bioinformatics, Supercomputing and Complex Genome Analysis*, 377-90.

Snyder, E. E. & Stormo, G. D. (1995). Identification of protein coding regions in genomic DNA. *J Mol Biol* **248,** 1-18.

Stolorz, P., Lapedes, A. & Xia, Y. (1992). Predicting protein secondary structure using neural net and statistical methods. *J Mol Biol* **225,** 363-77.

Tolstrup, N., Toftgard, J., Engelbrecht, J. & Brunak, S. (1994). Neural network model of the genetic code is strongly correlated to the GES scale of amino acid transfer free energies. *J Mol Biol* **243,** 816-20.

Van Ooyen, A. & Nienhuis, B. (1992). Improving the convergence of the back-propagation algorithm. *NEUNET* **5,** 465-71.

Vivarelli, F., Giusti, G., Villani, M., Campanini, R., Fariselli, P., Compiani, M. & Casadio, R. (1995). LGANN: a parallel system combining a local genetic algorithm and neural networks for the prediction of secondary structure of proteins. *Comput Appl Biosci* **11,** 253-60.

Wang, C., Venkatesh, S. S., and Judd, J. S. (1994). Optimal stopping and effective machine complexity in learning. *Adv Neural Inf Process Syst* **6,** 303-10.

Wu, C. & Shivakumar, S. (1994). Back-propagation and counter-propagation neural networks for phylogenetic classification of ribosomal RNA sequences. *Nucleic Acids Res* **22,** 4291-9.

Wu, C. H., Berry, M., Shivakumar, S. & McLarty, J. (1995). Neural networks for full-scale protein sequence classification: sequence encoding with singular value decomposition. *Machine Learning* **21,** 177-93.

Wu, C. H., Chen, H. L. & Chen, S. C. (1997). Counter-propagation neural networks for molecular sequence classification: Supervised LVQ and dynamic node allocation. *Applied Intelligence* **7,** 27-38.

CHAPTER 9

Applications - Nucleic Acid Sequence Analysis

9.1 Introduction

Some neural network applications for nucleic acid sequence analysis are summarized in Table 9.1, with an overview of their neural network designs and input/output encoding methods. The applications are discussed in three sections. The coding region recognition and gene identification problem (9.2) is tackled by two complementary approaches, gene search by signal and gene search by content (Staden, 1990). The *search by content* methods use various coding measures to determine the protein-coding potential of sequences. The *search by signal* methods identify signal sequences, such as splice sites, that delimit coding regions. Neural networks provide an attractive model in which sequence features for both signals and content can be combined and weighted to improve predictive accuracy (e.g., Uberbacher *et al.*, 1996; Snyder & Stormo, 1995).

The identification and analysis of other signals, binding sites or regulatory sites, such as promoters, ribosome-binding sites, and transcriptional initiating and terminating sites (9.3), are also important for the studies of gene regulation and expression. Common approaches to find functional signals include the consensus sequence method, the weight matrix method, and the neural network method. Neural networks allow the incorporation of both positive and negative examples as well as the detection of higher-order and long-range correlation, and are not based on the assumption of positional independence. As a result, neural networks are found to compare favorably to with other methods in many studies (e.g., Lapedes *et al.*, 1990; Bisant & Maizel, 1995; Larsen *et al.*, 1995).

Other applications of neural networks are sequence classification and feature detection (9.4). Detecting significant sequence features and understanding biological rules that govern gene structure and gene regulation are important problems that can be addressed by neural networks. Both supervised and unsupervised neural systems can be used as effective feature detectors. Features can be detected from within clusters of a self-organizing map (e.g., Arrigo *et al.*, 1991), or from the hidden units of a back-propagation multilayer perceptron (e.g., Tolstrup *et al.*, 1994). In addition to the back-propagation algorithm, effective neural classifiers can be trained using the adaptive resonance theory (LeBlanc *et al.*, 1994), or a counter-propagation algorithm (Wu *et al.*, 1994; 1997).

Table 9.1 *Neural network applications for nucleic acid sequence analysis.*

Reference	Application	Neural Network	I/O Encoding
Uberbacher & Mural, 1991	Coding Recognition	4L/FF/BP	Feat7/1(Y,N)
Uberbacher et al., 1996	Coding Recognition	3L/FF/BP	Feat13/1(Y,N)
Snyder & Stormo, 1993	I/E Feature Weighting	2L/FF/Delta	Feat6/1(Inequality)
Snyder & Stormo, 1995	I/E Feature Weighting	2,3L/FF/Delta,BP	Feat6/1(Inequality)
Brunak et al., 1991	Splice Site Prediction	3L/FF/BP	BIN4/1(Y,N)
Reese et al., 1997	Splice Site Prediction	3L/FF/BP	BIN4x2/1(Y,N)
Farber et al., 1992	Coding Recognition	2L/FF/BP	BIN4,Freq/1(Y,N)
Granjeon & Tarroux, 1995	I/E Composition Constraints	3L/FF/BP	BIN4/3(I,E,O)
Reczko et al., 1995	Parallel Coding Recognition	3L/FF/BP,QP,RP	BIN4/1(I,E)
Stormo et al., 1982	Ribosome Binding Site	Perceptron	BIN4/1(Y,N)
Bisant & Maizel, 1995	Ribosome Binding Site	3L/FF/BP	BIN4/1(Y,N)
Abremski et al., 1991	E. coli Promoter Prediction	3L/FF/BP	BIN4/1(Y,N)
Demeler & Zhou, 1991	E. coli Promoter Prediction	3L/FF/BP	BIN2,BIN4/1(Y,N)
O'Neill, 1991; 1992	E. coli Promoter Prediction	3L/FF/BP	BIN4/1(Y,N)
Horton & Kanehisa, 1992	E. coli Promoter Prediction	2L/FF/BP	BIN4+3+Freq/1(Y,N)
Mahadevan & Ghosh, 1994	E. coli Promoter Prediction	2x3L/FF/BP	BIN4/1(Y,N)
Pedersen & Engelbrecht,1995	Transcription Start Site	3L/FF/BP	BIN4/1(Y,N)
Pedersen& Nielsen, 1997	Translation Initiation Site	3L/FF/BP	BIN4/2(Y,N)
Larsen et al., 1995	Eukaryotic Promoter	3L/FF/BP	BIN4/1(Y,N)
Kraus et al., 1996	Eukaryotic Class II Promoter	2L/FF/BP	BIN4/1(Y,N)
Matis et al., 1996	RNA POL II Binding Site	4L/FF/BP	Feat13/1(Y,N)
Nair et al., 1994	Transcriptional Terminator	3L/FF/BP	BIN4,Real1/1(Y,N)
Nair et al., 1995	Transcription Control Signal	3L/FF/BP	BIN4/1(RTL)
Arrigo et al., 1991	Clustering & Function	2L/Kohonen	Real1/Map(30)
Giuliano et al., 1993	Clustering & Function	2L/Kohonen	Real1/Map
LeBlanc et al., 1994	Phylogenetic Classification	2L/ART	BIN4/19(Class)
Wu & Shivakumar, 1994	r-RNA Classification	2x3L/FF/BP,CP	Freq,SVD/220,15(Class)
Sun et al., 1995	t-RNA Gene Recognition	3L/FF/BP	BIN4/10(Class)
Tolstrup et al., 1994	Genetic Code Mapping	3L/FF/BP	BIN4/20(Class)

Neural network architectures: 2L/FF = two-layer, feed forward network (i.e., perceptron); 3L or 4L/FF = three- or four-layer, feed-forward network (i.e., multi-layer perceptron).

Neural network learning algorithms: BP = Back-Propagation; Delta = Delta Rule; QP = Quick-Propagation; RP = Rprop; ART = Adaptive Resonance Theory; CP = Counter-Propagation.

Input sequence encoding methods: BINn = binary-numbered direct encoding of residue identity, where n is the number of input units representing each residue; REALn = real-numbered direct encoding of residue features, where n is the number of units representing each residue; FEATn = indirect encoding of sequence features, where n is the number of units for the whole sequence or sequence window; FREQ = indirect encoding of sequence residue frequency; SVD = singular value decomposition; BIN4x2 = BIN4 encoding of dinucleotide.

Output sequence encoding methods, expressed by n(CODEs), where n is the number of output units, CODEs are: Y = Yes (positive); N = No (negative); I = Intron; E = Exon; O = Other (counter-example); RTL = relative transcription level.

9.2 Coding Region Recognition and Gene Identification

Pattern recognition methods play an important role in elucidating the locations and significance of genes throughout the genome. In prokaryotes, the coding region is a single open reading frame (ORF). Eukaryotic genes, however, are commonly organized as exons and introns and hence may comprise several disjoint ORFs. The main task of gene identification involves coding regions recognition (intron/exon discrimination) and splice sites detection.

Lapedes and colleagues (Lapedes *et al.*, 1990; Farber *et al.*, 1992) used a perceptron to predict coding regions in fixed-length windows with various input encoding methods, including the binary encoding (BIN4) of codon and the dicodon frequency. They concluded that the feature presentation using the dicodon frequency gave the best result, and that the neural network approach was superior to the Bayesian statistical prediction method that assumed independent codon frequency in each position.

The NetGene program (Brunak *et al.*, 1991) applies a three-layer back-propagation network and binary encoding method to predict acceptor and donor site positions in human genomic DNA sequences. It was observed that there exists a complementary relation between the strength of the splice site patterns and the confidence level of the coding/non-coding. Many weak splice sites have sharp transitions in the coding/no-coding signal, and vice versa. This correlation was exploited in NetGene, where a joint assignment, combining coding confidence level with the splice site strength, is used to reduce the number of false positives.

Granjeon and Tarroux (1995) studied the compositional constraints in introns and exons by using a three-layer network, a binary sequence representation, and three output units to train for intron, exon, and counter-example separately. They found that an efficient learning required a hidden layer, and demonstrated that neural network can detect introns if the counter-examples are preferentially random sequences, and can detect exons if the counter-examples are defined using the probabilities of the second-order Markov chains computed in junk DNA sequences.

Reczko *et al.* (1995) evaluated a parallel implementation of three-layer networks on a CM-5 machine with a MIMD (multiple instructions, multiple data) architecture. Several supervised learning algorithms were compared including back-propagation, *QuickProp* (Fahlman, 1988), and *Rprop* (Riedmiller & Braun, 1993). The best result was obtained with the Rprop training when BIN4 input of an 81-nucleotide window was mapped to a single output unit, determining whether the middle nucleotide belonged to an intron or exon.

More recently, many integrated programs for gene structure prediction and gene identification have been developed. Several of them, including GRAIL (Uberbacher *et al.*, 1996), GeneParser (Snyder & Stormo, 1995), and Genie (Kulp *et al.*, 1996; Reese *et al.*, 1997), have neural network components. The GRAIL system uses a neural network for coding region recognition. The original system (Uberbacher & Mural, 1991) used a four-layer back-propagation network to combine seven coding indicators calculated within a fixed sequence window. This network effectively weighted the various coding indicators on the basis of empirical data. The current system (Uberbacher *et al.*, 1996) considers discrete coding region candidates (of variable lengths) with specific edge signals, and incorporates additional indicators as inputs to a three-layer neural network. The total of 13 indicators includes splice site (donor/acceptor) strength, surrounding intron character, and candidate length.

The GeneParser system combines the neural network approach with the recursive optimization procedure of dynamic programming to predict gene structure. The neural network is used to weigh sequence information (both for sites and for content) and to approximate the log-likelihood that each subinterval exactly represents an intron or exon. A dynamic programming algorithm is then applied to this data to find the combination of introns and exons that maximizes the likelihood function. The neural network system is flexible in optimizing the GeneParser performance on sequences of atypical base composition or with simulated sequencing errors. Both perceptrons and three-layer networks were used (Snyder & Stormo, 1993; 1995), the former trained with a delta rule, the latter with back-propagation. Inputs to the networks were the differences for each statistic between the correct and incorrect solutions and the difference in the number of predicted sequence types. The output measured the inequality and was trained to a target value that maximized the difference between correct and incorrect solutions.

In the Genie program (Kulp *et al.*, 1996), a generalized hidden Markov model was used to describe the grammar of a legal parse of a DNA sequence. Probabilities were estimated for gene features by using dynamic programming to combine information from multiple content and signal sensors. The splice site sensors used two neural networks, one for the donor site and one for the acceptor site. The network design was as described in Brunak *et al.* (1991), where BIN4 encoding of input sequence windows was used.

Later, motivated by the observation that neighboring nucleotides at the splice sites were strongly correlated, Reese *et al.* (1997) replaced the two neural networks with dinucleotide encoding. Here, BIN4 encoding was applied to nucleotide pairs (i.e., bigrams), instead of individual nucleotides (i.e., monograms), with each indicator vector containing 15 zeros and a single one to indicate the identity of the corresponding nucleotide pair (e.g., AA). Hence, for an L-nucleotide long sequence window, the input vector size was 16 x (L-1) (i.e. L-1 bigrams), instead of 4 x L. The new input encoding approach improved both the sensitivity and specificity of gene structure identification, presumably due to its effectiveness in modeling pairwise correlations between adjacent nucleotides. The overall prediction accuracy was increased by approximately 5%.

9.3 Recognition of Transcriptional and Translational Signals

The tasks of transcriptional and translational signal recognition involve the prediction of promoters and sites that function in the initiation and termination of transcription and translation. Bacterial promoter sites, specifically the *Escherichia coli* RNA polymerase promoter site, are now very well characterized. The main problem is that the two conserved regions of the bacterial promoter, the -10 and -35 regions, are separated from each other by 15 to 21 bases, making the detection of the entire promoter as a single pattern difficult. Eukaryotic promoters are less well characterized than their bacterial equivalents. The major elements are the CCAAT box, GC box, TATA box and cap site.

The initiation codon, usually an AUG, signals the start of translation, and a termination codon marks the end of the translated region. In the analysis of prokaryotic DNA sequences, the signals include the transcriptional and translational initiation sites, the ribosome-binding site, and the transcriptional and translational termination sites. Due to the interrupted nature of the eukaryotic genes, the signals include the translation initiation sites, the intron/exon boundaries (splice sites), translational termination sites, and the polyadenylation sites.

Escherichia Coli Promoters

Prediction of promoter sequences using statistical and neural network approaches has shown that the simple consensus sequence prediction is not sufficient to identify putative promoters. Application of neural networks for the recognition of *E. coli* promoters is evident by the number of publications; many have been reviewed by Hirst and Sternberg (1992) and Presnell and Cohen (1993). With the exception of Lukashin *et al.* (1989), who used a simulated annealing approach for neural network training, most other early studies involved perceptrons or three-layer back-propagation networks (Abremski *et al.*, 1991; Demeler & Zhou, 1991; O'Neill, 1991; 1992) and used the BIN 4 representation of fixed-length windows ranging from 44 to 58 nucleotides.

Recent studies adopted more complex network architectures and/or input encoding. Horton and Kanehisa (1992) presented a data representation method that used seven single-position units to include the information of individual bases (BIN4) (4 units) and combination of two bases joined by an OR function (3 units). Another seven content units were used to include the content data (i.e., total counts) for each 12-nucleotide window. The total of 581 input units to the perceptron was trimmed by removing the input that was used the least (i.e., whose weight had the smallest absolute value). Mahadevan and Ghosh (1994) used a combination of two neural networks to identify *E. coli* promoters of all spacing classes (15 to 21 bases). The first neural network was used to predict the consensus boxes (-10 and -35, each containing 6 nucleotides); the second was designed to predict the entire sequence (65-nucleotide long) containing varying spacer lengths. Since the second neural net used the information of the entire sequence, possible dependencies existed between the bases in various positions. Poor training and prediction by a two-layer network without a hidden layer confirmed this possibility.

Pedersen and Engelbrecht (1995) devised a neural network to analyze *E. coli* promoters. They predicted the transcriptional start point, measured the information content, and identified new features signals correlated with the start site. They accomplished these tasks by using two different encoding schemes, one with windows of 1 to 51 nucleotides, the other with a 65-nucleotide window containing a 7-nucleotide hole. An interesting idea in the study was to measure the relative information content of the input data by using the ability of the neural network to learn correctly, as evaluated by the maximum test correlation coefficient.

Eukaryotic Promoters

Neural networks have also been applied to the analysis of eukaryotic promoters for RNA polymerase II. Larsen *et al.* (1995) used a three-layer back-propagation network to predict the exact location of the transcription initiation site in mammalian promoter regions. When long-range sequence windows (71 nucleotides) were used, at least two hidden units were needed to obtain significant positive values of the maximum correlation coefficient, indicating the nonlinearity of this feature space. Smaller windows were used to separate unimportant information from significant features. The best predictive result was obtained for a window containing both the TATA-box and the Cap signal using neural networks with hidden units.

Matis *et al.* (1996) applied a four-layer back-propagation network to predict the promoter scores using thirteen features including statistical (frequency) matrices and distance information, a design similar to the GRAIL neural network coding module. Kraus *et al.* (1996) used a perceptron to derive the weight matrices for the initiator (Inr) surrounding the transcription start site, and for the binding site for a TATA-binding protein (TBP) in class II (i.e., TATA-less) eukaryotic promoter. The investigators concluded that despite the lack of a consensus TATA box sequence, the -30 region element functions as a TBP binding site and cooperatively interacts with the Inr to determine the transcription site.

Ribosome-Binding Sites

Structural gene expression requires the binding of ribosomes to mRNA and the initiation of translation. In prokaryotes, the start codon is related to complementarity between the mRNA and the 16S ribosomal RNA (Shine & Dalgarno, 1975). The prediction of *E. coli* ribosome-binding sites (Shine-Dalgarno sequences) was the first neural network approach for molecular sequence analysis (Stormo *et al.*, 1982a). The study demonstrated the great promise of the neural network approach, although it utilized a simple perceptron architecture and a simple binary sequence encoding method. The perceptron, trained with the perceptron convergence theorem (Minsky & Papert, 1969), was more successful at finding the translational initiation sites than previous searches using rules (Stormo *et al.*, 1982b). Recently, Bisant and Maizel (1995) extended the study to include a larger data set using a three-layer network and concluded that the larger data gave better results and that a three-layer back-propagation network was much better than the perceptron on the same data.

Translation Initiation Sites

The prediction of translation initiation sites in eukaryotes involves determining which AUG triplet in an mRNA sequence is the start codon. Although translation usually starts at the first occurrence of the AUG triplet, downstream AUGs are used as start codons in less than 10% of reported eukaryotic mRNAs. The choice of the start codon is context dependent. A consensus sequence was derived with the analysis of 5'-noncoding sequences from a large number of vertebrate mRNAs (Kozak, 1987). Further analyses of sequences franking translational initiation sites for various eukaryotic taxonomic groups show that, although there are a few key similarities between taxonomic groups, a considerable degree of inter-taxon variation exists (Cavener & Ray, 1991).

Pedersen and Nielsen (1997) employed artificial neural networks to predict the initiation sites by distinguishing start and non-start AUGs using a combination of local start codon context and global sequence information. Two data sets were used: one from vertebrate, the other from a flowering plant, *Arabidopsis thaliana*. The system was a standard three-layer back-propagation network with BIN4 encoding and two output units to indicate whether the central position in the sequence window was the A in a start-codon or the A in a non-start codon. The results showed that the size of the input window was a significant factor for network performance, with better prediction from bigger windows, indicating the importance of global sequence information.

By selectively changing sequences in *E. coli* translation initiation region with randomized calliper inputs and observing the corresponding neural network performance, Nair (1997) analyzed the importance of the initiation codon and the Shine-Dalgarno sequence.

Transcription Terminator and Signal

Using 51-nucleotide sequence windows, Nair *et al.* (1994) devised a neural network to predict the prokaryotic transcription terminator that has no well-defined consensus patterns. In addition to the BIN4 representation (51 x 4 input units), an EIIP coding strategy was used to reflect the physical property (i.e., electron-ion interaction potential values) of the nucleotide base (51 units). The latter coding strategy reduced the input layer size and training time but provided similar prediction accuracy.

In another study, Nair *et al.* (1995) predicted the relative transcription levels of an eukaryotic globin gene and its use in the analysis of the transcriptional signals associated with the promoter region. Unlike most other applications, the output of the networks is a quantitative score, which measures the extent of gene expression. The simulation results can be used as a guide in designing mutational experiments.

9.4 Sequence Feature Analysis and Classification

DNA Feature Analysis

The Kohonen self-organizing map is useful for pattern clustering and feature detection. One of the first applications of the Kohonen map in molecular studies was conducted by Arrigo *et al.* (1991) to detect unique sequence features from DNA sequences. This application was made possible by the self-organizing network's ability to form internal representations that model the underlying structure of the input data. Analyses were performed without any previous alignment of sequence data. The coding values were computed by dividing the ordinal number of each base (A=1; C=2; G=3; T=4) by the corresponding molecular weight, which resulted in the base values: A=0.0374; C=0.0822, G=0.1059 and T=0.1683. After network training, a statistical analysis of the minimal similarity patterns (MSPs) determined the existence of singularity among the trained data. The statistical measure used was the Tanimoto similarity, which is directly proportional to the difference between each input vector and the weight vector. Giuliano *et al.* (1993) used a similar design to cluster genomic DNA sequences into MSP clusters. It is suggested that analysis of the clusters may help identify potentially functional DNA domains.

Phylogenetic Classification

LeBlanc *et al.* (1994) employed a hierarchical ART (Adaptive Resonance Theory), based on ART2 (Carpenter and Grossberg, 1991), to categorize tandem repeat DNA fragments from *Onchocerca volvulus*. The hierarchical ART network applied several ART2 networks to multiple layers to allow classification at different coarse levels, similar to a phylogenetic analysis. Each ART is a two-layer network with layer 1 (F1) for feature representation and layer 2 (F2) for category representation. The ART network can be used as a fast, easy-to-use classification tool that adapts to new data without retraining. When the network acquires new concepts, the old patterns are still kept in the memory; therefore, the network can learn incrementally without reviewing old instances. This nice feature is missing in supervised learning networks. However, the ability of ART networks to generalize is limited by the lack of a hidden layer.

Wu and Shivakumar (1994) developed a neural network system for classification of ribosomal RNAs according to phylogenetic classes. Two separate networks were trained, one for 220 small subunit classes, and the other for 15 large subunit classes. The input sequences were encoded using various n-gram encoding schemes, followed by a singular value decomposition compression to different number of reduced dimensions. It was found that the first 100 orthogonal components of the 65,536-dimensional input vector (from 8-gram frequencies) were sufficient for the best result. Two learning algorithms were used, the back-propagation and a modified counter-propagation algorithm with supervised LVQ (learning vector quantizer) and dynamic node allocation (Wu *et al.*, 1997). The counter-propagation networks were trained rapidly, about one to two orders of magnitude faster than back-propagation networks. When combining the results from the back-propagation and counter-propagation networks, the predictive accuracy approached 100%. The program is part of the gene classification artificial neural system (GenCANS) available for on-line search (Wu, 1996).

Sun *et al.* (1995) applied a three-layer back-propagation network to classify transfer RNA gene sequences according to their source organisms. The evolutionary relationship derived from this study was consistent with those from other methods.

Genetic Code Mapping

A unique neural network application (Tolstrup *et al.*, 1994) was to map the DNA genetic code (BIN4 representation of codons) into amino acid categories (20 output units). This mapping illustrates an important use of neural networks for a problem that is completely data driven. No a priori relationship between the nucleotides or amino acids is introduced. Another interesting feature of this neural network design is its use of a minimal network to capture the internal representation of the input-output mapping. Here, the hidden units act essentially as feature detectors. In contrast to many other neural networks, the weights of the trained network have a fairly comprehensive structure. The analysis of the two hidden units, which contained internal representation of genetic code mapping, show that the category assignment was strongly correlated to an amino acid transfer free energy scale.

9.5 References

Abremski, K., Sirotkin, K. & Lapedes, A. (1991). Application of neural networks and information theory to the identification of *E.coli* transcriptional promoters. *Math Model Sci Comput* **2**, 634-41.

Arrigo, P., Giuliano, F., Scalia, F., Rapallo, A. & Damiani, G. (1991). Identification of a new motif on nucleic acid sequence data using Kohonen's self organizing map. *Comput Appl Biosci* **7**, 353-7.

Bisant, D. & Maizel, J. (1995). Identification of ribosome binding sites in Escherichia coli using neural network models. *Nucleic Acids Res* **23**, 1632-9.

Brunak, S., Engelbrecht, J. & Knudsen, S. (1991). Prediction of human mRNA donor and acceptor sites from the DNA sequence. *J Mol Biol* **220**, 49-65.

Carpenter, G. A. & Grossberg, S. (eds.). (1991). *Pattern Recognition by Self-Organizing Neural Networks*. MIT P, Cambridge.

Cavener, D.R. & Ray, S. C. (1991). Eukaryotic start and stop translation sites. *Nucleic Acids Res* **19**, 3185-92.

Demeler, B. & Zhou, G. W. (1991). Neural network optimization for E. coli promoter prediction. *Nucleic Acids Res* **19**, 1593-9.

Fahlman, S. E. (1988) Faster-learning variations on backprogagation: an empirical study. In *Proceedings of the Connectionist Models Summer School* (ed. Touretzky, D., Hinton, G. and Sejnowski, T.). Morgan Kaufman, San Mateo, CA.

Fahlman, S., and Lebiere, C. (1990). The cascade-correlation learning architecture. *Adv Neural Inf Process Syst* **2**, 524-32.

Farber, R., Lapedes, A. & Sirotkin, K. (1992). Determination of eukaryotic protein coding regions using neural networks and information theory. *J Mol Biol* **226**, 471-9.

Giuliano, F., Arrigo, P., Scalia, F., Cardo, P. P. & Daminani, G. (1993). Potentially functional regions of nucleic acids recognized by a Kohonen's self-organizing map. *Comput Appl Biosci* **9**, 687-93.

Granjeon, E. & Tarroux, P. (1995). Detection of compositional constraints in nucleic acid sequences using neural networks. *Comput Appl Biosci* **11**, 29-37.

Hirst, J. D. & Sternberg, M. J. (1992). Prediction of structural and functional features of protein and nucleic acid sequences by artificial neural networks. *Biochemistry* **31**, 7211-8.

Horton, P. B. & Kanehisa, M. (1992). An assessment of neural network and statistical approaches for prediction of E. coli promoter sites. *Nucleic Acids Res* **20**, 4331-8.

Kozak, M. (1987). An analysis of 5'-noncoding sequences from 699 vertebrate messenger RNAs. *Nucleic Acids Res* **15**, 8125-48.

Kraus, R. J., Murray, E. E., Wiley, S. R., Zink, N. M., Loritz, K., Gelembiuk, G. W. & Mertz, J. E. (1996). Experimentally determined weight matrix definitions of the initiator and TBP binding site elements of promoters. *Nucleic Acids Res* **24**, 1531-9.

Kulp, D., Haussler, D., Reese, M. G. & Eeckman, F. H. (1996). A generalized hidden Markov model for the recognition of human genes in DNA. *Ismb* **4**, 134-42.

Lapedes, A., Barnes, C., Burks, C., Farber, R. & Sirotkin, K. (1989). Application of neural networks and other machine learning algorithms to DNA sequence analysis. In: *Computers and DNA, SFI Studies in the Sciences of Complexity*, vol. 7 (ed. Bell, G. I. & Marr, T. G.), pp. 157-82. Addison-Wesley, Rosewood City, CA.

Larsen, N. I., Engelbrecht, J. & Brunak, S. (1995). Analysis of eukaryotic promoter sequences reveals a systematically occurring CT-signal. *Nucleic Acids Res* **23**, 1223-30.

LeBlanc, C., Katholi, C. R., Unnasch, T. R. & Hruska, S. I. (1994). DNA sequence analysis using hierarchical ART-based Classification Networks. *Ismb* **2**, 253-60.

Lukashin, A. V., Anshelevich, V. V., Amirikyan, B. R., Gragerov, A. I. & Frank-Kamenetskii, M. D. (1989). Neural network models for promoter recognition. *J Biomol Struct Dyn* **6**, 1123-33.

Mahadevan, I. & Ghosh, I. (1994). Analysis of E.coli promoter structures using neural networks. *Nucleic Acids Res* **22**, 2158-65.

Matis, S., Xu, Y., Shah, M., Guan, X., Einstein, J. R., Mural, R. & Uberbacher, E. (1996). Detection of RNA polymerase II promoters and polyadenylation sites in human DNA sequence. *Comput Chem* **20**, 135-40.

Minsky, M. & Papert, S. (1969). *Perceptrons*. MIT P, Cambridge.

Nair, T. M., Tambe, S. S. & Kulkarni, B. D. (1994). Application of artificial neural networks for prokaryotic transcription terminator prediction. *FEBS Lett* **346**, 273-7.

Nair, T. M., Tambe, S. S. & Kulkarni, B. D. (1995). Analysis of transcription control signals using artificial neural networks. *Comput Appl Biosci* **11**, 293-300.

Nair, T. M. (1997). Calliper randomization: an artificial neural network based analysis of *E. coli* ribosome binding sites. *J Biomol Struct Dyn* **15**, 611-7.

O'Neill, M. C. (1991). Training back-propagation neural networks to define and detect DNA-binding sites. *Nucleic Acids Res* **19**, 313-8.

O'Neill, M. C. (1992). Escherichia coli promoters: neural networks develop distinct descriptions in learning to search for promoters of different spacing classes. *Nucleic Acids Res* **20**, 3471-7.

Pedersen, A. G. & Engelbrecht, J. (1995). Investigations of *Escherichia coli* promoter sequences with artificial neural networks: new signals discovered upstream of the transcriptional startpoint. *Ismb* **3**, 292-9.

Pedersen, A. G. & Nielsen, H. (1997). Neural network prediction of translation initiation sites in eukaryotes: perspectives for EST and genome analysis. *Ismb* **5**, 226-33.

Presnell, S. R. & Cohen, F. E. (1993). Artificial neural networks for pattern recognition in biochemical sequences. *Annu Rev Biophys Biomol Struct* **22**, 283-98.

Reczko, M., Hatzigeorgiou, A., Mache, N., Zell, A. & Suhai, S. (1995). A parallel neural network simulator on the connection machine CM-5. *Comput Appl Biosci* **11**, 309-15.

Reese, M. G., Eeckman, F. H., Kulp, D. & Haussler, D. (1997). Improved splice site detection in Genie. *J Comput Biol* **4**, 311-23.

Riedmiller, M. & Braun, H. (1993). A direct adaptive method for faster backpropagation learning: the Rprop algorithm. In *Proceedings of the IEEE International Conference on Neural Networks (ICNN 93)*. (ed. Ruspini H.), pp. 586-91.

Shine, J. & Dalgarno, L. (1975). Terminal-sequence analysis of bacterial ribosomal RNA. Correlation between the 3'-terminal-polypyrimidine sequence of 16-S RNA and translational specificity of the ribosome. *Eur J Biochem* **57**, 221-30.

Snyder, E. E. & Stormo, G. D. (1993). Identification of coding regions in genomic DNA sequences: an application of dynamic programming and neural networks. *Nucleic Acids Res* **21**, 607-13.

Snyder, E. E. & Stormo, G. D. (1995). Identification of protein coding regions in genomic DNA. *J Mol Biol* **248**, 1-18.

Staden, R. (1990). Protein coding regions in genomic sequences. Methods Enzymol **183**, 163-80.

Stormo, G. D., Schneider, T. D., Gold, L. & Ehrenfeucht, A. (1982a). Use of the 'Perceptron' algorithm to distinguish translational initiation sites in E. coli. *Nucleic Acids Res* **10**, 2997-3011.

Stormo, G. D., Schneider, T. D. & Gold, L. M. (1982b). Characterization of translational initiation sites in E. coli. *Nucleic Acids Res* **10**, 2971-96.

Sun, J., Song, W. Y., Zhu, L. H. & Chen, R. S. (1995). Analysis of tRNA gene sequences by neural network. *J Comput Biol* **2**, 409-16.

Tolstrup, N., Toftgard, J., Engelbrecht, J. & Brunak, S. (1994). Neural network model of the genetic code is strongly correlated to the GES scale of amino acid transfer free energies. *J Mol Biol* **243**, 816-20.

Uberbacher, E. C. & Mural, R. J. (1991). Locating protein-coding regions in human DNA sequences by a multiple sensor-neural network approach. *Proc Natl Acad Sci U S A* **88**, 11261-5.

Uberbacher, E. C., Xu, Y. & Mural, R. J. (1996). Discovering and understanding genes in human DNA sequence using GRAIL. *Methods Enzymol* **266**, 259-81.

Wu, C. & Shivakumar, S. (1994). Back-propagation and counter-propagation neural networks for phylogenetic classification of ribosomal RNA sequences. *Nucleic Acids Res* **22**, 4291-9.

Wu, C. H. (1996). Gene Classification Artificial Neural System. *Methods Enzymol* **266**, 71-88.

Wu, C. H., Chen, H. L. & Chen, S. C. (1997). Counter-propagation neural networks for molecular sequence classification: supervised LVQ and dynamic node allocation. *Applied Intelligence* **7,** 27-38.

CHAPTER 10

Applications - Protein Structure Prediction

Table 10.1 *Neural network applications for protein structure prediction.*

Reference	Application	Neural Network	Input/Output Encoding
Qian & Sejnowski, 1988	2-D Structure	2,3L/FF/BP	BIN21/3(α,β,C)
Bohr *et al.*, 1988	2-D Structure	2L/FF/BP	BIN20/2(Y,N)
Holley & Karplus, 1989	2-D Structure	3L/FF/BP	BIN21/2(α,β,C)
Kneller *et al.*, 1990	2-D Structure	2L/FF/BP	BIN21(α,β,C)
Stolorz *et al.*, 1992	2-D Structure	2,3L/FF/BP	BIN21/3(α,β,C)
Sasagawa & Tajima, 1993	2-D Structure	3L/FF/BP	BIN24/2(α,b,C) or 1(Y,N)
Fariselli *et al.*, 1993	Membrane 2-D Structure	2,3L/FF/BP	BIN20/3 or 4(α,β,C,τ)
Rost & Sander, 1993a; b	2-D Structure	2x3L/FF/BP	Prf21;Real3/3(α,β,L)
Rost & Sander, 1994a	2-D Structure	2x3L/FF/BP	Prf21+Real3+%AA/3(α,β,L)
			Real4+Real1+%AA/3(α,β,L)
Vivarelli *et al.*, 1995	2-D Structure	2,3L(GA)/FF/BP	BIN30/3(α,β,C)
Chandonia & Karplus,1995	2-D Structure	3L/FF/BP	BIN21/2(α,β)
Rost *et al.*, 1995	Transmembrane Helices	2x3L/FF/BP	Prf21+Real3+%AA/2(Y,N)
			Real3+Real1+%AA/2(Y,N)
Fariselli & Casadio, 1996	Transmembrane Helices	2x3L/FF/BP	BIN20;Real2/2(Y,N)
Aloy *et al.*, 1997	Transmembrane Helices	3L/FF/BP	BIN21x21/1(Y,N)(1,-1)
Rost & Sander, 1994b	Solvent Accessibility	3L/FF/BP	Prf21+Real3+%AA/10(Level)
Muskal & Kim, 1992	2-D Structure Content	2x3L/FF/BP	%AA+Feat;MF/2(α,β)
Sun *et al.*, 1997	Super2-D Structure	3L/FF/BP	Real1/11(Class)
Bohr *et al.*, 1990	3-D Backbone Structure	3L/FF/BP	BIN20/30(dist)(α, β,C)
Xin *et al.*, 1993	3-D Structure	3L/FF/BP	Real1/140x140 (dist)
Milik *et al.*, 1995	3-D Side-Chain Packing	3L/FF/BP	BIN(7x7map)/1(Y,N)
Dosztanyi *et al.*, 1997	Stabilization Center	2,3L/FF/BP	BIN21x17 & Prf21x17/2 (Y,N)
Jones, 1999	Fold Alignment	3L/FF/BP	Feat6/2(Y/N)
Lund *et al.*, 1997	Distance Constraints	3L/FF/BP	BIN20+BIN20/1(Y/N)
Dubchak *et al.*, 1993a; b	Folding Classes	2,3L/FF/BP	%AA+Feat/1-4(Class)
Dubchak *et al.*, 1995	Folding Classes	3L/FF/BP	Feat/2(Y/N)
Chandonia & Karplus,1995	Structural Classes	3L/FF/BP	%AA+Feat/4(Class)
Chandonia & Karplus,1996	Structural Classes	3L/FF/SCG	%AA+Feat/4(Class)

Neural network architectures and learning algorithms: (see Table 9.1); GA = Genetic Algorithm; SCG = Scaled Conjugate Gradient.

Input/Output sequence encoding methods: (see Table 9.1); PRF =direct encoding of residue profile; α = α-helix; β = β-sheet; C = random coil; t = β-reverse turn; L = loop; %AA = amino acid composition; FEAT = indirect encoding of sequence features; MF = memory factor; dist = distance; Map = Kohonen map (with dimensions).

10.1 Introduction

Table 10.1 summarizes neural network applications for protein structure prediction. Protein secondary structure prediction is often used as the first step toward understanding and predicting tertiary structure because secondary structure elements constitute the building blocks of the folding units. An estimated 90% or so of the residues in most proteins are involved in three classes of secondary structures, the α-helices, β-strands or reverse turns. Related to the secondary structure prediction are also the prediction of solvent accessibility, transmembrane helices, and secondary structure content (10.2). Neural networks have also been applied to protein tertiary structure prediction, such as the prediction of the backbones or side-chain packing, and to structural class prediction (10.3).

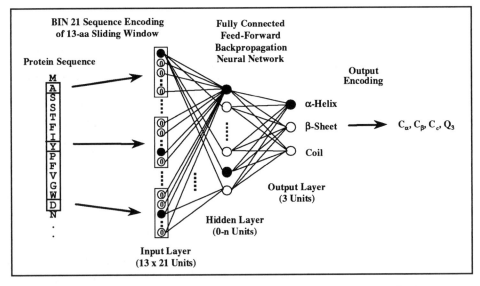

Figure 10.1 *Neural network for protein secondary structure prediction.*
(Adopted from Qian & Sejnowski, 1988).

10.2 Protein Secondary Structure Prediction

Protein secondary structure prediction is one of the earliest neural network applications in molecular biology, and has been extensively reviewed. Typified by the work of Qian and Sejnowski (1988) (Figure 10.1), early studies involved the use of perceptron or three-layer back-propagation networks and the binary (BIN) sequence encoding method (Bohr *et al.*, 1988; Holley & Karplus, 1989; Kneller *et al.*, 1990; Stolorz *et al.*, 1992; Sasagawa & Tajima, 1993; Fariselli *et al.*, 1993). Minor variations to the BIN21 scheme were used in two studies. Kneller *et al.* (1990) added one additional input unit to present the hydrophobicity scale (a real number) of each amino acid residue, and showed slightly increased accuracy. Sasagawa and Tajima (1993) used BIN24 to encode three additional amino acid alphabets, B, X, and Z.

Protein sequences were processed as sliding windows of fixed-length segments, usually ranging from 7 to 17 amino acids. The central residues were then given either a 3-state (α helix, β sheet, and random coil) or 2-state (e.g., helix, non-helix) prediction. The 3-state prediction was encoded using three (e.g., 100 for helix, 010 for sheet, 001 for coil) or two output units (e.g., 10 for helix, 01 for sheet, 00 for coil). The perceptron weight matrices were analyzed using a Hinton diagram to show the relative contribution to the secondary structure made by each amino acid at each position (e.g., Qian & Sejnowski, 1988; Kneller *et al.*, 1990).

Since these studies used local encoding schemes which utilized limited correlation information between residues, little or no improvement was shown by using a multi-layered network with hidden units (Qian & Sejnowski, 1988; Stolorz *et al.*, 1992; Fariselli *et al.*; 1993). A performance ceiling of about 65% three-state accuracy was observed in these networks. The results were only marginally more accurate than a simplistic Bayesian statistical method that assumed independent probabilities of amino acid residues (Stolorz *et al.*, 1992).

It has been observed that the use of protein tertiary structural class improved the accuracy for a 2-state secondary structure prediction (Kneller *et al.*, 1990). A modular network architecture was proposed using separate networks (i.e., α- or β-type network) for classification of different secondary structures (Sasagawa & Tajima, 1993). Recently, Chandonia & Karplus (1995) trained a pair of neural networks to predict the protein secondary structure and the structural class respectively. Using predicted class information, the secondary structure prediction network realized a small increase in accuracy.

Fariselli *et al.* (1993) extended the neural network design for secondary structure prediction of globular proteins to the prediction of membrane proteins. The result was better than those obtained with statistical methods used for membrane proteins. Their findings also indicated that regular patterns of secondary structures are common for globular and membrane proteins.

Vivarelli *et al.* (1995) used a hybrid system that combined a local genetic algorithm (LGA) and neural networks for the protein secondary structure prediction. The LGA, a version of the genetic algorithms (GAs), was particularly suitable for parallel computational architectures. Although the LGA was effective in selecting different neural network topologies, it was inferior to back-propagation as a neural network learning algorithm. The results showed that several neural network topologies can similarly perform on the testing set, independent of their adjustable parameters. The prediction result, based on single input sequences and local encoding scheme, was similar to those of earlier studies (e.g., Qian & Sejnowski, 1988). This finding indicates that the uppermost limit for the accuracy weakly depends on the specific network architecture, and confirms the relevance of the input information as a determining factor.

The accuracy level of predicting protein secondary structure using the neural network approach has been recently substantially increased, surpassing a 70% level of the average three-state accuracy (Rost & Sander, 1993; Rost, 1996). Although the basic neural network design remained the same by using the three-layer back-propagation networks, there were several important improvements. These included the use of evolutionary information, the use of cascaded neural networks, and the incorporation of global information (Figure 10.2).

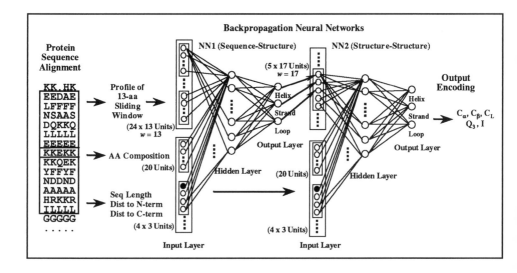

Figure 10.2 *Neural network for protein secondary structure prediction.*
(Adopted from Rost & Sander, 1993; Rost, 1996).

At the first level, the sequence-to-structure network (NN1) was trained to classify mutually independent segments of residues (13-amino acid-long) in terms of the state of a single central residue. A new key aspect was the use of evolutionary information contained in multiple sequence alignments as input to a neural network in place of single sequences. Each sequence position was represented by the amino acid residue frequencies derived from the HSSP (Sander & Schneider, 1991) alignments (Rost & Sander, 1993), or by a profile generated from a set of homologous sequences (Rost, 1996). A total of 24 real-valued input units was used for each residue. The input vector consisted of 20 units for its sequence profile, one for the spacer (allowing the window to extend over the N- and C-terminal ends), one for the position-specific conservation weight, and two for the numbers of insertions and deletions for the position. In addition, global information was incorporated, including the amino acid composition of the protein (20 real numbers), as well as the sequence length and the distances of the central amino acid to the protein ends (4 binary numbers each to code for values in four ranges). The total number of input units, thus, is 24 x 13 (for $w = 13$) + 20 + 12.

A second level structure-to-structure network (NN2) was used to introduce a correlation between the secondary structure of adjacent residues in 17-amino acid windows. The structure context training provided better prediction of helix and strand lengths. The output (the three-state values of the central residue) from the first network was input to the second network. In addition, a spacer unit and a conservation weight unit were used for each residue, as well as the global information of the protein. The total number of input units for NN2 is 5 x 17 (for $w = 17$) + 20 + 12. The secondary structure prediction program (PHDsec) is available via an automatic E-mail prediction service (Rost, 1996).

Although the overall accuracy of secondary structure prediction is significantly improved with the PHDsec design, not all proteins can be equally well predicted. Worst predicted proteins are those with unusual features and those with bad sequence alignments. When information about homologous proteins is not available (i.e., no multiple sequence alignment), the predictive accuracy may be reduced by about 10%.

Solvent Accessibility

Rost and Sander (1994b) developed another neural network system to predict the relative solvent accessibility (PHDacc). The one-level network system used the same input information as that in the PHDsec sequence-to-structure network, and mapped it to ten output units coded for ten relative levels of solvent accessibility. PHDacc was superior to other methods in predicting the residues in either of the two states, buried or exposed. Entirely buried residues (<4% accessible) were predicted best.

The accuracy of secondary structure and residue solvent accessibility predictions using Rost and Sander's neural networks has been recently evaluated with a newly determined X-ray crystal structure, the C-terminal half of the heavy chain of tetanus toxin (Lebeda *et al.*, 1998). Based on the standardized accuracy measures such as Q, C, and I (Chapter 8.5), the network-derived predictions are compared with the observed secondary structures as determined by the DSSP program. Because the accuracies of these predictions are comparable to those made by Rost and Sander using a dataset of 126 nonhomologous globular proteins, the study provides a quantitative foundation for gauging the results when building by homology the structures of related proteins.

Helical Transmembrane Domain Prediction

Rost *et al.* (1995) developed a neural network system (PHDhtm) to predict the location of transmembrane helices (HTM) based on evolutionary information. The design was similar to the one used for prediction of protein secondary structure for globular proteins that involved two cascaded neural networks. Same-input information including sequence profiles and global information was provided to the sequence-to-structure network. The output from the first network as well as global information became input to the second structure-to-structure network. Both neural networks had two output units, coded for HTM and not-HTM. The method significantly improved the overall per-residue accuracy of predicting residues in HTM, and could warrant use of the predictions as a starting point for a complete *ab initio* prediction of tertiary structure for transmembrane regions. The method was also applied to yeast genomic sequence analysis.

Fariselli and Casadio (1996) described a program for predicting the location and orientation of α-helical transmembrane segments in integral membrane proteins. The program used two cascaded multi-layer networks with single protein sequences encoded by BIN20 method as input. The neural network output was the protein membrane topology based on the statistical propensity of residues. The system was capable of predicting non-homologous proteins with high accuracy.

Secondary Structure Content

A novel neural network design involving tandem networks was used to predict the protein secondary structure content and to address the issue of network memorization (Muskal & Kim, 1992). Two neural nets were used, NET1 to predict the structure content and NET2 to predict NET1's state of memorization. The input to NET1 was the amino acid composition of the protein (20 units), the molecular weight (1 unit), and the presence or absence of the heme group (1 unit). The output was the helix and strand content of the protein (2 units). The input to NET2 was the memory factor calculated for each of the H hidden nodes in NET1 (H units), and the number of network training iterations. The output was the memory state for the helix and strand prediction (2 units). NET2 was then used to determine when NET1 was in a state of generalization and was able to give the best structure composition prediction. Comparative studies revealed that the neural network approach was better than a multiple linear regression method and a two-layer perceptron.

Protein Supersecondary Structure

High resolution X-ray analysis of protein structures shows that the conformational categories of the connecting peptides which link the α-helices and β-sheets are limited. Such well defined types of folding units, such as αα- and ββ-hairpins, and αβ- and βα-arches, are referred to as supersecondary structures. One important step towards building a tertiary structure from secondary structures is to identify these supersecondary structure building blocks. Sun *et al.* (1997) employed three-layer back-propagation neural networks to predict supersecondary structures from protein sequences. The training set, which was compiled from a supersecondary structure motif database (Sun & Jiang, 1996), contained 11 types of frequently occurring motifs. Each motif consisted of two regular secondary structures (α-helix or β-sheet) and a connecting peptide with five types of linking residues (a, b, l, e, and t) in the coil conformation.

The network input was the supersecondary motif sequence encoded by an ordinal number (ranging from 1 to 20) arbitrarily assigned to each amino acid (e.g., Ala=6; Asn= 4). Thus, the size of the input vector was the same as the length of the sequence window. The output vector had 11 units, representing the 11 supersecondary motifs. Eleven individual neural networks were trained, each having only one output unit turned on to represent the designated secondary motif. In the prediction phase, the test sequence was assigned to the motif category of the winning network with the largest output value. The results showed a more than 70% predictive accuracy.

10.3 Protein Tertiary Structure Prediction

Protein Tertiary Structure

The earliest neural network attempt for protein tertiary structure prediction was done by Bohr *et al.* (1990). They predicted the binary distance constraints for the C-α atoms in protein backbone using a standard three-layer back-propagation network and BIN20 sequence encoding method for 61-amino acid windows. The output layer had 33 units, three for the 3-state secondary structure prediction, and the remaining to measure the distance constraints between the central amino acid and the 30 preceding residues.

Wilcox and colleagues (Wilcox *et al.*, 1991; Xin *et al.*, 1993) applied a large-scale neural network to learn the protein tertiary structures in the Protein Data Bank represented by 140 x 140 distant matrices. The sequence-structure mapping was accomplished by encoding the entire protein sequence (66-129 residues in length) into 140 input units. The amino acid residue was represented by its hydrophobicity scale, normalized to a real value of between -1 and 1. The resulting network produced good predictions of distance matrices from homologous sequences, but had limited generalization capability due to the small size of the training set relative to the network size.

Milik *et al.* (1995) developed a neural network system to evaluate side-chain packing in protein structures. Instead of using protein sequence as input to the neural network as in most other studies, protein structure represented by a side-chain-side-chain contact map was used. Contact maps of globular protein structures in the Protein Data Bank were scanned using 7 x 7 windows, and converted to 49 binary numbers for the neural network input. One output unit was used to determine whether the contact pattern is popular in the structure database. The procedure may be used to determine the restriction of the conformational space and to detect regions where supersecondary structure packing is non-protein-like.

Stabilization Center

Dosztanyi *et al.* (1997) utilized neural networks to predict residues in stabilization center elements, which consist of clusters of residues making cooperative long-range interactions. A stabilization center is identified in protein contact maps where an accumulation of long-range interactions is observed. Standard back-propagation neural networks with and without a hidden layer were developed using 19-amino acid windows of single sequences (BIN21 encoding) or sequence alignments (PRF21 encoding) as input. Although perceptrons and three-layered networks yielded similar results, the incorporation of evolutionary information from homologous proteins improved predictive accuracy. The stabilization center residues had distinct composition; relative accessibility, number and type of interactions; and a higher structural and sequential conservation than the randomized reference set.

Fold Recognition and Alignment Evaluation

Fold recognition method has become an important approach to the protein structure prediction problem (Chapter 1.1.2). One particular method for fold recognition, GenTHREADER (Jones, 1999), has a neural network component to produce a single measure of confidence in the proposed prediction. The method is divided into three stages: alignment of sequences, calculation of pair potential and solvation terms, and evaluation of the alignment using a neural network. Fold recognition methods commonly generate a number of scores which relate to different aspects of the sequence-structure alignment. In GenTHREADER, six scores were deemed important: initial sequence profile alignment score, number of aligned residues, length of target sequence, length of template protein sequence, pairwise energy sum, and solvation energy sum.

Due to the interdependent and nonlinear nature of these scores, a three-layer neural network was employed to weigh the scores and evaluate sequence-structure alignment quality. The network architecture consisted of six input units for each of the scores scaled to the range of 0-1 by using the standard logistic function, and two output units to determine whether the proteins are related or unrelated based on a threshold value. The importance of and relationship between different input scores were studied by analyzing the effect of variable input combinations on the network output. Although the network output ranged between 0 and 1, it did not directly imply a probability value. Instead, the confidence level of the prediction was estimated based on the false positive rates at different output values. The method was applied to the analysis of a *Mycoplasma* genome and showed that more than 45% of the proteins derived from the predicted protein coding regions have a significant relationship to a protein of known structure.

Protein Distance Constraints

While both homology modeling and threading are powerful methods for protein structure prediction, they rely on the presence of a similar structure. One general method of studying protein folding is to generate protein distance constraints and apply these in algorithms that perform structure refinement, threading, or approximation of fold structure (e.g., Skolnick *et al.*, 1997). Lund *et al.* (1997) devised neural networks to predict interatomic C^α distances and compared the results with probability density functions. Three-layer back-propagation networks were used to predict whether distances between a particular pair of amino acids, at a given sequence separation, were above or below a given distance threshold.

The input consisted of two sequence windows, each containing 9 or 15 amino acids and separated by different lengths (i.e., sequence separation). It was observed that the optimal window size was 18 (i.e., two 9-amino acid windows) for short sequence separations, and 30 for large sequence separations. One output unit was used to indicate contact (above distance threshold) and non-contact (below threshold) between the central amino acids of the two sequence windows. To predict distance inequalities for whole proteins, multiple networks were trained, each on a different sequence separation. In order to assess the significance of the neural network output quantitatively, the training set was used to establish a relationship between the network output o and the probability p that the amino acid pair was closer than a given threshold. Using this table to convert the network output to probabilities, it was shown that about 62% of the inequalities in the test set were correctly predicted. The predictive results were better than a method based on pair density functions (i.e., probability distributions of the pairwise distance between two amino acids) which are statistically derived from the training set. With the use of sequence windows in the input, the neural network approach provided context information of the amino acid residues not available otherwise.

10.4 Protein Folding Class Prediction

One approach to the protein structure prediction is to classify the folding patterns of globular proteins. This is based on the observation from examining known tertiary structures that the variety of protein folding patterns is significantly restricted. Therefore, it is likely that a protein may belong to one of the previously identified folding patterns.

Various schemes have been developed for the classification of protein three-dimensional structures. One common scheme is the classification based on the four *tertiary super classes*, namely, all α (proteins having mainly α-helix secondary structure), all β (mainly β-sheet secondary structure), α+β (segment of α-helices followed by segment of β-sheets), and α/β (alternating or mixed α-helix and β-sheet segments) (Levitt, 1976). A fifth class is often added to account for globular proteins with irregular secondary structural arrangements. Some other classification schemes consider macro folding patterns, such as the globin fold, the Rossmann fold, and the antiparallel β barrels; or folding motifs, such as the Greek key, EF-hand, and Kringle motifs. In the database of structural classification of proteins, SCOP (Hubbard *et al.*, 1999), comprehensive protein classifications are provided on hierarchical levels according to their structures and evolutionary origin (Murzin *et al.*, 1995). Similarly, the CATH database (Orengo *et al.*, 1999) provides a hierarchical classification of protein domain structures into evolutionary families and structural groupings, and currently has 593 fold groups.

Dubchak *et al.* (1993a; 1993b) used perceptrons and three-layer networks to predict several protein folding classes from the amino acid composition. The protein sequence was converted to neural network input using two representations. One used the length and amino acid composition of the protein (21 units), the other reduced the number of input (7 units) by using physical subdivision (i.e., hydrophobic composition characteristics) of amino acids into a few classes. The output was one to four folding classes, for 4α-helical bundles, parallel $(\alpha/\beta)_8$ barrels, nucleotide binding fold and immunoglobulin fold, respectively. An integrated neural network package, PROBE (Holbrook *et al.*, 1993), was developed to make predictions of the four folding motifs and compare them with the secondary structure prediction methods (Qian & Sejnowski, 1988; Holley & Karplus, 1989) as well as the overall secondary structure composition (Muskal & Kim, 1992).

Later, Dubchak *et al.* (1995) extended the three-layer neural networks to predict 83 folding classes of 254 proteins in the 3D_ali database (Pascarella & Argos, 1992; Pascarella *et al.*, 1996). The input sequence encoding involved the use of global protein sequence descriptors for the major physicochemical amino acid attributes. The amino acid attributes were the relative hydrophobicity of amino acids (hydrophobic, neutral, and polar), predicted secondary structure (helix, strand, and coil), consensus secondary structure resulted from two predictions (helix, strand, coil, and unknown), and predicted solvent accessibility (buried and exposed). The global descriptors were the composition, transition, and distribution, to describe the composition of a given amino acid property (i.e., alternative alphabet in Chapter 6.6), the frequencies with which the properties changes along the entire length of the protein, and the distribution pattern of the property along the sequence, respectively. For each of the amino acid attribute, all three descriptors were combined and used as input to the neural nets. By using the various combination of descriptors, individual networks were trained, one for each of the 83 classes with two output units to distinguish this class from all other classes. It was expected that additional descriptors and input units would be needed in order to assign a protein sequence to an even larger number of folding classes or to further improve the network performance.

When the similar approach was applied to discriminate members of a given folding class from members of all other classes in the more comprehensive SCOP database, it was shown that specific amino acid properties work differently on different folding classes

(Dubchak *et al.*, 1997). Therefore, one may find an individual set of descriptors that works best on a particular folding class.

The basic information of protein tertiary structural class can help improve the accuracy of secondary structure prediction (Kneller *et al.*, 1990). Chandonia and Karplus (1995) showed that information obtained from a secondary structure prediction algorithm can be used to improve the accuracy for structural class prediction. The input layer had 26 units coded for the amino acid composition of the protein (20 units), the sequence length (1 unit), and characteristics of the protein (5 units) predicted by a separate secondary structure neural network. The secondary structure characteristics include the predicted percent helix and sheet, the percentage of strong helix and sheet predictions, and the predicted number of alterations between helix and sheet. The output layer had four units, one for each of the tertiary super classes (all-α, all-β, α/β, and other). The inclusion of the single-sequence secondary structure predictions improved the class prediction for non-homologous proteins significantly by more than 11%, from a predictive accuracy of 62.3% to 73.9%.

Later, a similar study was conducted (Chandonia & Karplus, 1996) using the larger data set of 318 non-homologous chains (Sali & Overington, 1994), which was roughly five times the size of the set of 62 protein chains used in the above study. The neural networks were trained using a scaled conjugate gradient algorithm (Moller, 1993), which ran about an order of magnitude faster than the back-propagation algorithm and scaled better for larger networks. With the larger data set and the introduction of additional hidden units (resulting in a final architecture of 26 input units, 9 hidden units and 4 output units), the structural class prediction accuracy was improved from 73.9% in the previous study to 80.2%.

10.5 References

Aloy, P., Cedano, J., Oliva, B., Aviles, F. X. & Querol, E. (1997). 'TransMem': a neural network implemented in Excel spreadsheets for predicting transmembrane domains of proteins. *Comput Appl Biosci* **13**, 231-4.

Bohr, H., Bohr, J., Brunak, S., Cotterill, R. M., Lautrup, B., Norskov, L., Olsen, O. H. & Petersen, S. B. (1988). Protein secondary structure and homology by neural networks. The alpha- helices in rhodopsin. *FEBS Lett* **241**, 223-8.

Bohr, H., Bohr, J., Brunak, S., Cotterill, R. M. J., Fredholm, H., et al. (1990). A novel approach to prediction of the 3-dimensional structures of protein backbones by neural networks. *FEBS Lett* **261**, 43-6.

Chandonia, J. M. & Karplus, M. (1995). Neural networks for secondary structure and structural class predictions. *Protein Sci* **4**, 275-85.

Chandonia, J. M. & Karplus, M. (1996). The importance of larger data sets for protein secondary structure prediction with neural networks. *Protein Sci* **5**, 768-74.

Dosztanyi, Z., Fiser, A. & Simon, I. (1997). Stabilization centers in proteins: identification, characterization and predictions. *J Mol Biol* **272**, 597-612.

Dubchak, I., Holbrook, S. R. & Kim, S.-H. (1993a). Prediction of protein folding class from amino acid composition. *Proteins* **16**, 79-91.

Dubchak, I., Holbrook, S. R. & Kim, S.-H. (1993b). Comparison of two variations of neural network approach to the prediction of protein folding pattern. *Ismb* **1**, 118-26.

Dubchak, I., Muchnik, I., Holbrook, S. R. & Kim, S. H. (1995). Prediction of protein folding class using global description of amino acid sequence. *Proc Natl Acad Sci U S A* 92, 8700-4.

Dubchak, I., Muchnik, I. & Kim, S. H. (1997). Protein folding class predictor for SCOP: approach based on global descriptors. *Ismb* **5**, 104-7.

Fariselli, P., Compiani, M. & Casadio, R. (1993). Predicting secondary structure of membrane proteins with neural networks. *Eur Biophys J* **22**, 41-51.

Fariselli, P. & Casadio, R. (1996). HTP: a neural network-based method for predicting the topology of helical transmembrane domains in proteins. *Comput Appl Biosci* **12**, 41-8.

Holbrook, S. R., Dubchak, I. & Kim, S.-H. (1993). PROBE: A computer program employing an integrated neural network approach to protein structure prediction. *Biotechniques* **14**, 984-9.

Holley, H. L. & Karplus, M. (1989). Protein secondary structure prediction with a neural network. *Proc Natl Acad Sci USA*. **86**, 152-6.

Hubbard, T. J., Ailey, B., Brenner, S. E., Murzin, A. G. & Chothia, C. (1999). SCOP: a Structural Classification of Proteins database. *Nucleic Acids Res* **27**, 254-6.0

Jones, D. T. (1999). GenTHREADER: An efficient and reliable protein fold recognition method for genomic sequences. *J Mol Biol* **287**, 797-815.

Kabsch, W. & Sander, C. (1983). Dictionary of protein secondary structure: pattern recognition of hydrogen-bonded and geometrical features. *Biopolymers* **22**, 2577-637.

Kneller, D. G., Cohen, F. E. & Langridge, R. (1990). Improvements in protein secondary structure prediction by an enhanced neural network. *J Mol Biol* **214**, 171-82.

Lebeda, F. J., Umland, T. C., Sax, M. & Olson, M. A. (1998). Accuracy of secondary structure and solvent accessibility predictions for a clostridial neurotoxin C-fragment. *J Protein Chem* **17**, 311-8.

Levitt, M. & Chothia, C. (1976). Structural patterns in globular proteins. *Nature* **261**, 552-8.

Lund, O., Frimand, K., Gorodkin, J., Bohr, H., Bohr, J., Hansen, J. & Brunak, S. (1997). Protein distance constraints predicted by neural networks and probability density functions. *Protein Eng* **10**, 1241-8.

Milik, M., Kolinski, A. & Skolnick, J. (1995). Neural network system for the evaluation of side-chain packing in protein structures. *Protein Eng* **8**, 225-36.

Moller, M. (1993). A scaled conjugate gradient algorithm for fast supervised learning. *Neural Networks* **6**, 525-33.

Murzin, A. G., Brenner, S. E., Hubbard, T. & Chothia, C. (1995). SCOP: a structural classification of proteins database for the investigation of sequences and structures. *J Mol Biol* **247**, 536-40.

Muskal, S. M. & Kim, S.-H. (1992). Predicting protein secondary structure content. A tandem neural network approach. *J Mol Biol* **225**, 713-27.

Orengo, C. A., Pearl, F. M., Bray, J. E., Todd, A. E., Martin, A. C., Lo Conte, L. & Thornton, J. M. (1999). The CATH Database provides insights into protein structure/function relationships. *Nucleic Acids Res* **27**, 275-9.

Pascarella, S & Argos, P. (1992). A data bank merging related protein structures and sequences. *Protein Eng* **5**, 121-37.

Pascarella, S., Milpetz, F. & Argos, P. (1996). A databank (3D-ali) collecting related protein sequences and structures. *Protein Eng* **9,** 249-51.

Qian, N. & Sejnowski, T. J. (1988). Predicting the secondary structure of globular proteins using neural network models. *J Mol Biol* **202,** 865-84.

Rost, B. & Sander, C. (1993a). Improved prediction of protein secondary structure by use of sequence profiles and neural networks. *Proc Natl Acad Sci U S A* **90,** 7558-62.

Rost, B. & Sander, C. (1993b). Prediction of protein secondary structure at better than 70% accuracy. *J Mol Biol* **232,** 584-99.

Rost, B. & Sander, C. (1994a). Combining evolutionary information and neural networks to predict protein secondary structure. *Proteins* **19,** 55-72.

Rost, B. & Sander, C. (1994b). Conservation and prediction of solvent accessibility in protein families. *Proteins* **20,** 216-26.

Rost, B., Casadio, R., Fariselli, P. & Sander, C. (1995). Transmembrane helices predicted at 95% accuracy. *Protein Sci* **4,** 521-33.

Rost, B. (1996). PHD: predicting one-dimensional protein structure by profile-based neural networks. *Methods Enzymol* **266,** 525-39.

Sali, A & Overington, J. P (1994). Derivation of rules for comparative protein modeling from a database of protein structure alignments. *Protein Sci* 3, 1582-96.

Sander, C. & Schneider, R. (1991). Database of homology-derived protein structures and the structural meaning of sequence alignment. *Proteins* **9,** 56-68.

Sasagawa, F. & Tajima, K. (1993). Prediction of protein secondary structures by a neural network. *Comput Appl Biosci* **9,** 147-52.

Skolnick, J., Kolinski, A. & Ortiz, A. R. (1997). MONSSTER: a method for folding globular proteins with a small number of distance restraints. *J Mol Biol* **265,** 217-41.

Stolorz, P., Lapedes, A. & Xia, Y. (1992). Predicting protein secondary structure using neural net and statistical methods. *J Mol Biol* **225,** 363-77.

Sun, Z. & Jiang, B. (1996). Patterns and conformations of commonly occurring supersecondary structures (basic motifs) in protein data bank. *J Protein Chem* **15,** 675-90.

Sun, Z., Rao, X., Peng, L. & Xu, D. (1997). Prediction of protein supersecondary structures based on the artificial neural network method. *Protein Eng* **10,** 763-9.

Vivarelli, F., Giusti, G., Villani, M., Campanini, R., Fariselli, P., Compiani, M. & Casadio, R. (1995). LGANN: a parallel system combining a local genetic algorithm and neural networks for the prediction of secondary structure of proteins. *Comput Appl Biosci* **11,** 253-60.

Wilcox, G. L., Poliac, M. O. & Liebman, M. N. (1991). Neural network analysis of protein tertiary structure. *Tetrahedron Comp Meth* **3,** 191-211.

Xin, Y., Carmeli, T. T., Liebman, M. N. & Wilcox, G. L. (1993). Use of the backpropagation neural network algorithm for prediction of protein folding patterns. In *Proceedings of the Second International Conference on Bioinformatics, Supercomputing, and Complex Genome Analysis* (ed. Lim, H. A., Fickett, J. W., Cantor, C. R. & Robbins, R. J.), pp. 359-75. World Scientific, New Jersey.

CHAPTER 11

Applications - Protein Sequence Analysis

11.1 Introduction

Sequence similarity database searching and protein sequence analysis constitute one of the most important computational approaches to understanding protein structure and function. Although most computational methods used for nucleic acid sequence analysis are also applicable to protein sequence studies, how to capture the enriched features of amino acid alphabets (Chapter 6) poses a special challenge for protein analysis.

Table 11.1 Neural network applications for protein sequence analysis.

Reference	Application	Neural Network	Input/Output Encoding
Ladunga et al., 1991	Signal Peptide Prediction	4L/FF/Tiling	BIN20/1(Y,N)
Nielsen et al., 1997	Signal Peptide Prediction	2x3L/FF/BP	BIN21/1(Y,N)
Schneider et al.,1994;1995	Cleavage Site Feature	2L/FF/GA	Real2/1(Y,N)
Wrede et al., 1998	Site Feature/Peptide Design	2L/FF/GA	Real2/1(Y,N)
Hansen et al., 1995; 1998	O-Glycosylation Site	3L/FF/BP	BIN21/1(Y,N)
Blom et al., 1996	Cleavage Site	3L/FF/BP	BIN21/1(Y,N)
Lohmann et al., 1994	Membrane-Spanning Region	4L/FF/GA	Real7/1(Y,N)
Nakata, 1995	Zinc Finger Motif	2,3L/BP	BIN(Vary)/1(Y,N)
Gulukota et al., 1997	MHC-Binding Motif	3L/FF/BP	BIN20/1(Y,N)
Ferran & Ferran, 1992	Family Classification	2L/Kohonen	Freq/Map(7x7)
Ferran & Pflugfelder,1993	Family Classification	2L/Kohonen	Freq,PCA/Map(7x4)
Ferran et al., 1994	Family Classification	2L/Kohonen	Freq/Map(15x15)
Wu et al., 1992	Family Classification	3L/FF/BP	Freq/137-178(Class)
Wu et al., 1995	Family Classification	3L/FF/BP	Freq,SVD/167-416(Class)
Wu et al., 1996	Family/Motif Identification	3L/FF/BP	Freq/4(G,M)

Neural network architectures and learning algorithms: (see Table 9.1); GA = genetic Algorithm; Tiling = Tiling algorithm.

Input/Output sequence encoding methods: (see Table 9.1); PCA = Principal Component Analysis; Map = Kohonen map (with dimensions); G = Global (member/non-member); M = Motif (member/non-member).

Neural network applications for protein sequence analysis are summarized in Table 11.1. Like the DNA coding region recognition problem, signal peptide prediction (11.2) involves both search for content and search for signal tasks. An effective means for protein sequence analysis is *reverse database searching* to detect functional motifs or sites (11.3) and identify protein families (11.4). Most of the functional motifs are also units of folding motifs. Generally the sequence-structure relationship is not strong enough to enable prediction, but short motifs with a very high sequence signal and thus high predictability may play an important role in folding (Rooman *et al.*, 1992). Such fragments tend to be preserved in evolution more than other fragments within protein families, suggesting that they have an important role in folding.

11.2 Signal Peptide Prediction

The biosynthesis of many proteins requires their transport across the endoplasmic reticulum membrane, in eukaryotes, or the cytoplasmic membrane, in prokaryotes. *Signal peptides* are essential for targeting nascent protein chains to translocation sites in the membrane and are cleaved off from mature proteins. Evidence shows that signal sequences are exchangeable between different classes of organisms, and work in much the same way in both eukaryotes and prokaryotes (Gierasch, 1989; Rapoport, 1992). Despite this striking conservation of a critical cellular function, signal sequences display a remarkable lack of sequence similarity, even among closely related proteins. Thus, the recognition of signal peptides represents a challenge to the traditional computational sequence analysis methods that infer functions primarily based on sequence similarity. There is, however, a strong industrial interest in signal sequences for finding more effective vehicles to produce proteins in recombinant systems. Sequence-oriented *de novo* design of cleavage sites and functional peptides is now possible with the advances in recombinant DNA technology and heterologous expression systems.

Although no clear sequence similarity exists, known signal peptides do share some defining characteristics and have stronger sequence conservation at the *cleavage site* where they are recognized by signal pepidase. Rules have been deduced from signal sequences of various proteins to describe the common features required for cleavage specificity (von Heijne, 1983; von Heijne, 1985). Signal peptides are short amino-terminal segments that generally comprise a positively charged amino-terminal region (*n-region*) of 2 to 5 residues, a central hydrophobic region (*h-region*) of 7 to 15 residues, and a polar carboxy-terminal cleavage region (*c-region*) of 5 to 7 residues. The cleavage site has a more specific pattern, especially at positions -3 and -1, as described by the *-3, -1 rule* to contain only small and uncharged residues. Further analyses of signal peptides from a larger collection of prokaryotic and eukaryotic species have shown variations in lengths and amino acid compositions and distinct differences among species in all three regions (von Heijne & Abrahmsen, 1989; Nielsen *et al.*, 1997).

Signal peptide identification, like DNA intron/exon sequence discrimination, involves the two related problems of *signal peptide discrimination* (search for content) and *cleavage site recognition* (search for signal). It is well suited to neural network methods for several reasons. The functional units are encoded by local, linear sequences of amino acids rather than global 3-dimensional structures (Claros *et al.*, 1997). The ambiguity of sequence conservation requires that sequence discrimination be studied both from a chemical and a biochemical perspective (von Heijne, 1995). The search for content is further complicated by the wide length and compositional variations.

Due to the stronger sequence conservation (signal) around the cleavage site, the weight matrix method can be used for site recognition (von Heijne, 1986). The predictive accuracy is estimated to be around 75-80% for both prokaryotic and eukaryotic proteins. With the length and compositional variations, the search of signal peptides based on their common features along the sequences (content), however, cannot be resolved using the weight matrix method. Ladunga *et al.* (1991) designed a neural network classifier to identify whether each of the 20-amino acid peptides was an amino-terminal signal peptide or a part of a cytosolic protein. A *tiling algorithm* (Mezard and Nadal, 1989) was used to optimize the network topology by adding hidden units and layers dynamically and resulted in a final network of two hidden layers. When the neural network classifier was used together with von Heijne's weight matrix method (1986), an improvement over either individual method alone was observed, as would be expected due to the combined search for both signal and content.

Recently, Nielsen *et al.* (1997) have developed a comprehensive system for signal peptide prediction that combines the search for signal and content. Pairs of neural networks were trained for sequence discrimination and site detection separately, and tailored to different sets of prokaryotic and eukaryotic sequence. The data sets were compiled from the SwissProt protein database with special care to remove redundant (highly homologous) sequences and to avoid entries whose sites are not experimentally determined (Nielsen *et al.*, 1996). The data was organized according to their source organisms into different sets, which include human, *E. coli*, eukaryotes (including human), Gram-negative bacteria (including *E. coli*), and Gram-positive bacteria, but exclude the *Mycobacteria* and archaes (the third kingdom besides eukaryotes and prokaryotes). The division of the data sets conceivably can help adjust for the marked differences (especially length) observed between signal peptides from different types of organisms.

The pairs of neural networks, which were used to discern signal and non-signal sequences and to detect the cleavage site, adopted the same sequence encoding and neural network design. Amino acid residues in sequence windows of varying lengths were represented by indicator vectors (BIN21) and trained by back-propagation neural networks of various hidden sizes. The network output was the probability of the central amino acid belonging to a signal sequence or being the cleavage site (S-score and C-score). An overall measure (Y-score) was then computed by combining the two scores. After performance evaluation with cross validation, best network was configured for each data set with a different window size and number of hidden units. Direct comparisons between the weight matrix and neural network methods using the same data sets showed that the two methods were comparable for cleavage site prediction. However, the neural network method performs significantly better than previous prediction schemes when both S- and

C-scores are considered. The system has been applied to complete genomic sequence analysis and is currently considered as the standard method for signal peptide prediction. On-line prediction is available from the WWW server at http://www.cbs.dtu.dk.

Cleavage Site Feature Extraction and Peptide Design

A novel approach was developed by Schneider and colleagues to analyze the cleavage site for the purpose of finding optimized cleavage sites for protein export. Their studies included the signal peptidase I cleavage sites in *E. coli* (Schneider *et al.*, 1994), and the mitochondrial matrix metalloprotease target sequence in *N. crasa* (Schneider *et al.*, 1995). In order to extract the physicochemical features of the amino acid residues, continuous scales of two properties, volume and hydrophobicity, were used for sequence description. The two properties are orthogonal and appropriate for feature extraction of the signal peptide site, considering the characteristic hydrophobic h-region and the size constraints indicated in the *-3, -1 rule*. Sequence windows of 13-amino acids covering positions −10 to +2 (of the mature protein) from cleavage and non-cleavage sites were used for network training.

A two-layer perceptron with a sigmoidal transfer function for the network's output unit was used to reflect the functional relationship between the physiochemical properties of the cleavage-site sequence and their efficiency for protein export. The perceptron was trained using an evolution strategy (Lohmann *et al.*, 1994). The linear combination of outputs of multiple perceptrons was used as a neural filter to provide the fitness function for an SME (*simulated molecular evolution*) procedure to select optimal sites. The idea of the SME is similar to that of the *genetic algorithms* (Goldberg, 1989; Koza, 1992; Holland, 1992). The SME design started from a random sequence, generated new sequences by a mutation procedure, made selection of sequences based on their quality evaluated by the neural filter system, and stopped after certain number of optimization cycles (generations) or when a high quality sequence had been generated.

In their recent study, Wrede *et al.* (1998) confirmed that the artificial cleavage site sequences generated from the neural network approach was active *in vivo* in an *E. coli* expression and secretion system. Upon comparison of the natural and artificial cleavage sites, it was observed that although a proline residue is predominant in position −6 of natural signal peptidase I cleavage sites, glycine was selected by the SME. It was also observed that the positions −2 and −6 of SME-generated sequences were comparably restricted in the variability of the tolerated residues. These and other comparative analysis results suggest that the neural network model used in the design process may have considered possible interactions and dependencies between residues, and recognized crucial site features that are more complex than just residue frequencies at individual positions.

11.3 Other Motif Region and Site Prediction

Neural networks have been used successfully for the detection of binding or regulatory sites in nucleic acid sequences that lack clear consensus sequences. This is also true for the prediction of cleavage or acceptor sites in protein sequences.

Glycosylation Sites

Glycosylation serves a wide variety of functions in biology. Almost all the secreted and membrane-associated proteins of eukaryotic cells are glycosylated. Oligosaccharides form two types of direct attachments to these proteins: N-linked and O-linked. N-glycosylation occurs exclusively on the nitrogen atom of Asn side chains, whereas O-glycosylation occurs on the oxygen atoms of hydroxyls, particularly those of Ser and Thr residues. The prediction of N-linked glycosylation sites is straightforward, as the enzyme specificity can be formulated as a consensus triplet sequence. On the other hand, there are several classes of O-linked glycosylation. Type-specific enzymes are involved in transferring sugars to specific acceptor residues, none of which exhibits a clear consensus sequence pattern. Studies have been conducted to predict the most common type of O-glycosylation site, the mucin-type. The goal was to predict, from protein sequence alone, whether specific Ser and Thr residues in a glycoprotein are O-glycosylated and linked to an N-acetylgalactosamine (GalNAc) sugar moiety.

From the analyses of many glycosylation sites (GalNAc transferase acceptor sites), a few rules of thumb have been formulated (Wilson *et al.*, 1991) and motif patterns have been proposed (Pisano *et al.*, 1993). Matrix statistics have been compiled for site prediction (Elhammer *et al.*, 1993). However, since there is no clear consensus acceptor sequence pattern and it is strongly influenced by the local conformation, neural network is appropriate for the task.

Hansen *et al.* (1995) used feed-forward back-propagation neural networks with BIN21 encoding to determine the position of O-linked GalNAc-glycosylated Ser and Thr residues. The neural network approach was much more reliable than a weight matrix method and any other available method. The weight matrix method used the sequence context (relative frequency of amino acids) around glycosylated residues only. The neural networks, on the other hand, allow the use of information in both glycosylated (positive) and non-glycosylated (negative) sites. The neural network was also effective in detecting correlations at a longer range and at sites with nonlinear feature space.

Later, Hensen *et al.* (1998) extended the work by using a jury of four differently trained glycosylation neural networks and one surface accessibility network. The surface accessibility network was used to derive a modulated threshold (i.e., cutoff value for glycosylation network output) because O-glycosylation sites were found exclusively on the surface of proteins. If the site and surroundings were predicted surface accessible, the cutoff was lowered. The resulting system correctly identified 83% glycosylated and 90% non-glycosylated Ser and Thr residues.

Other Cleavage Sites

Blom *et al.* (1996) applied a similar neural network design to distinguish cleavage from noncleavage sites in picornaviral polyproteins. A logarithmic error function was used in place of the conventional RMS error function during network training. It reduced the convergence time considerably, and has the property of making a given network architecture learn more complex tasks without increasing the network size. The performance of the networks approached 100% accuracy.

Membrane-Spanning Region Prediction

Highly related to the signal peptide recognition problem is that of the membrane-spanning region. The most common prediction method is the analysis of hydrophobicity profiles. The hydrophobic core of signal peptides or helical structures can give rise to false positive results. Most techniques are not suitable for membrane proteins composed of β-sheets. Lohmann *et al.* (1994; 1996) developed a hybrid neural network system for the prediction of membrane-spanning sequences. A set of 47 human integral membrane proteins containing 162 experimentally determined transmembrane regions were used as positive examples. Negative examples consisting of sequence windows not covering the membrane/non-membrane border were used at a ratio of 4:1 to positive examples. Among the 47 membrane proteins, 11 randomly selected proteins were used as the prediction set (with 48 positive examples), and the remaining were separated into three training sets (with 38 positive examples each). An evolutionary algorithm (i.e., genetic algorithm) was used to optimize the neural network topology into a four-layer partially-connected feed-forward network and to train the network. The results were similar to those obtained with the hydrophobicity plot. Since the training sequences used in the neural system adopted helical structures within the membrane, the system focused on the recognition of transmembrane helical structures only.

In these studies, the input sequence was represented by seven non-orthogonal physicochemical properties, namely, hydrophobicity, volume, surface area, hydrophilicity, bulkiness, refractivity, and polarity, with normalized values of (-1, 1). Thirteen-amino acid sequence windows were used, resulting in an input vector of 91 (i.e., 13 x 7) units. With the small number of training examples, the input units were only partially connected to the hidden layer in order to reduce the number of free parameters and avoid overtraining the network. The constraint was to connect each first hidden layer unit to one amino acid property exclusively, and that the size of the second hidden layer size was no larger than the first hidden layer. The output layer had exactly one unit to indicate whether the middle residue was in a membrane/non-membrane boarder. The process of development started with randomly generated architectures and resulted in 18 and 8 units in the first and second hidden layers, with a total number of free parameters of

175. Considering convergence properties of the evolutionary algorithm by comparing the number of different structures occurred in the process (about 655,000) to the size of the stock of structures (10^{70}), it was suggested that local search was carried out in the network architecture and parameter spaces.

The neural network architecture optimized by the evolutionary algorithm could be analyzed for a biochemical interpretation and feature extraction. One may infer the importance of input properties based on the relative connectivity of the input units. For example, bulkiness, which was not connected at all, was probably unimportant. On the other hand, units for polarity, refractivity, hydrophobicity, and surface area were highly connected, indicating these are important features of membrane transition regions.

Zinc Finger Motifs

Nakata (1995) used back-propagation networks to predict zinc finger DNA binding proteins by analyzing the characteristic patterns around the zinc finger motifs. In the motif, the zinc atom is bound to four amino acids with some combination of cysteines (Cys) and histidines (His). Two zinc finger types were studied. The TFIIIA type contains an antiparallel β-ribbon and an α-helix, with two invariant Cys, two invariant His, and three conserved hydrophobic residues. The steroid hormone receptor type has two consecutive zinc fingers, each having four invariant Cys and a few conserved residues, and bind to DNA as dimers. Sequence motifs of 31 to 37-amino acids and 64 to 77-amino acids were used for training the two zinc finger types, respectively.

The input sequences were presented by the BIN encoding of various alphabets that extract different chemicophysical and structural properties of amino acids, such as the charge and polarity, hydrophobicity, and secondary structure propensity. Features of both single amino acid residues and residue pairs were encoded. Since the BIN (direct) encoding required fixed-length sequence windows, the shorter motif fragments were filled with blanks at training positions. Two-layer perceptrons were used to map individual features, whereas combinations of features were trained using three-layer perceptrons. Predictive accuracy of the individual neural networks was then analyzed to identify the most effective features for sequence discrimination. The neural network predictive results were similar to those of discriminant analysis using combinations of attributes, reaching an accuracy of 96 to 97% for predicting both types of zinc fingers.

Major Histocompatibility Complex Binding Motif

The prediction of peptides which bind to major histocompatibility complex (MHC) molecules is important for identifying autoimmune and CTL epitopes and for peptide vaccine design because the MHC-binding is a necessary condition for the peptide to be a T-cell epitope. Although MHC-binding peptides exhibit certain common sequence motifs, they are neither necessary nor sufficient for binding. Many binding peptides do not have the characteristic motifs and less than one third of motif-containing peptides actually bind.

Gulukota *et al* (1997) developed a neural network to predict the MHC-binding peptides and compared its performance with a polynomial method and the standard motif-based method. Both of the latter methods assumed independent binding of individual side-chains. The data set consisted of 463 nonamer peptides with known binding affinity (IC_{50}) to HLA-A2.1, 273 of which carried the nominal motif (i.e., L/M/I at position 2 and L/V/I at position 9). The neural networks used were three-layer back-propagation nets consisting of 180 input units (BIN20 encoding of the 9-amino acid peptide sequence) and one output unit (for a binary prediction of binding or non-binding). In contrast to the neural network method, which had no motif requirement for its training set, the polynomial method was trained using only peptides containing the canonical motif. The results showed that the two methods were complementary and both were superior to sequence motifs.

11.4 Protein Family Classification

There are two different approaches to the protein sequence classification problem. One can use an unsupervised neural network to group proteins if there is no knowledge of the number and composition of final clusters (e.g., Ferran & Ferrara, 1992). Or one can use supervised networks to classify sequences into known (existing) protein families (e.g., Wu *et al.*, 1992).

Ferran and his group used self-organizing Kohonen maps for protein family clustering and phylogenetic classification. The original network mapped the bi-peptide amino acid frequency of protein sequences (i.e., 400 input units) into a 7 x 7 feature map of protein family clusters (Ferran & Ferrara, 1992). In a subsequent study, Ferran and Pflugfelder (1993) reduced the number of inputs from 400 into 20 principal components using the principal component analysis. The feature map was configured to be 7 x 4 according to a statistical clustering analysis. The neural network-statistics hybrid system resulted in a much smaller network and faster training, with similar results. Ferran *et al.* (1994) scaled up the system to classify all human protein sequences using a 15 x 15 feature map. The bi-peptide frequency of a reduced alphabet of 11 letters (i.e., 11 amino acid groups) was as effective as the original 20-letter amino acid alphabet. The neural network approach was shown to be better than statistical non-hierarchical clustering method on this complex data.

Wu *et al.* (1992) devised a neural network system for the automatic classification of protein sequences according to superfamilies. It was extended into a full-scale system for classification of more than 3,300 PIR protein superfamilies (Wu *et al.*, 1995). The basic input information was encoded with the n-gram and SVD methods. The implementation of the full system adopted a modular network architecture that involved multiple independent neural networks to partition different protein functional groups. During the training phase, each of the 13 network modules was trained separately using the sequences of known superfamilies; during the prediction phase, unknown sequences were classified on all modules with classification results combined.

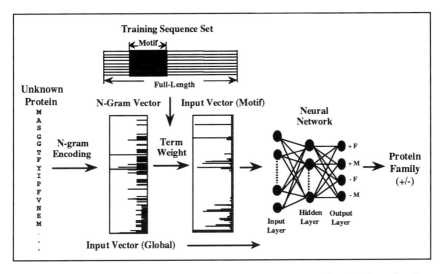

Figure 11.1 MOTIFIND neural network design for protein family classification.

The design was enhanced for the detection of distant family relationships through the use of both global and motif sequence information (Wu *et al.*, 1996). The motif identification neural design (MOTIFIND) (Figure 11.1) used an n-gram term weighting algorithm (Figure 11.2) to extract conserved family information from local sequence motifs. The global input vector concatenated the bi-grams of amino acids (A2) (20^2 units) and tetra-grams of exchange groups (E4) (6^4 units). A weighting factor derived from sequence motifs for each individual n-gram was used to compute the motif input vector. With the A2E4 sequence encoding, both global and motif vectors had a vector size of 1696 (i.e., $20^2 + 6^4$). Individual three-layered feed-forward back-propagation neural networks were trained, one for each protein family. Each network had four output units to code for the global and motif pattern of both positive (member) and negative (non-member) classes. The global score was calculated from full-length network outputs from positive classes and the motif score was calculated from motif network output for positive classes. The final network architecture was 1696 x 20 x 4.

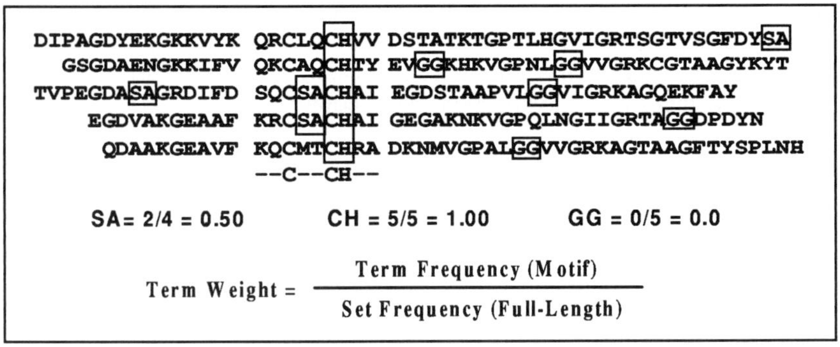

DIPAGDYEKGKKVYK QRCLQCHVV DSTATKTGPTLHGVIGRTSGTVSGFDYSA
GSGDAENGKKIFV QKCAQCHTY EVGGKHKVGPNLGGVVGRKCGTAAGYKYT
TVPEGDASAGRDIFD SQCSACHAI EGDSTAAPVLGGVIGRKAGQEKFAY
EGDVAKGEAAF KRCSACHAI GEGAKNKVGPQLNGIIGRTAGGDPDYN
QDAAKGEAVF KQCMTCHRA DKNMVGPALGGVVGRKAGTAAGFTYSPLNH
--C--CH--

SA= 2/4 = 0.50   CH = 5/5 = 1.00   GG = 0/5 = 0.0

$$\text{Term Weight} = \frac{\text{Term Frequency (Motif)}}{\text{Set Frequency (Full-Length)}}$$

Figure 11.2 N-Gram term weighting method to extract motif information.

Integrated System for Protein Family Classification

Protein family relationships can be studied according to the extent of sequence similarity at the whole protein, domain, and motif levels, as mentioned before (Chapter 1.1.3). An ideal family classification system, therefore, should provide information regarding both global and local sequence similarity and their relevance to protein functions. An integrated system, GeneFIND gene family identification network design (Wu *et al.*, 1999a) (Figure 11.3), was developed based on the MOTIFIND neural networks and the ProClass family database (Wu *et al.*, 1999b) for database searching against protein families. The design objectives were to improve the speed and sensitivity for family identification, to differentiate global and motif similarities, and to provide family information for assisting functional annotation of proteins.

As a full-scale family classification system, more than 1200 MOTIFIND neural networks were implemented, one for each ProSite protein group. The training set for the neural networks consisted of both positive (ProSite family members) and negative (randomly selected non-members) sequences at a ratio of 1 to 2. ProClass groups non-redundant SwissProt and PIR protein sequence entries into families as defined collectively by PIR superfamilies and ProSite patterns. By joining global and motif similarities in a single classification scheme, ProClass helps to reveal domain and family relationships, and classify multi-domained proteins.

GeneFIND uses a multi-level filter system, with MOTIFIND and BLAST (Altschul *et al.*, 1997) as the first-level filters to quickly eliminate query sequences that have very low probabilities of being a family member. After searching through all neural networks, the sequence query is considered as a potential PROSITE family member if it is ranked in the top 3% hits of the corresponding network.

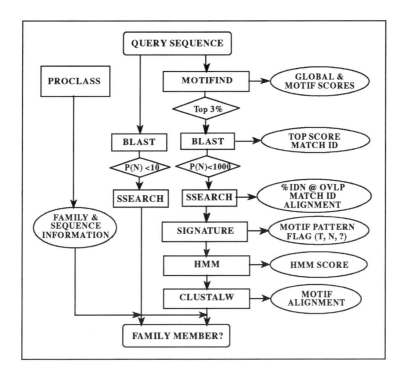

Figure 11.3 *GeneFIND family identification system.*

Potential member detected by MOTIFIND and BLAST is further searched against each member of the corresponding family using the Smith-Waterman (1981) SSEARCH program (Pearson, 1991) to detect the best pair-wise alignment. For possible ProSite family members, motif pattern matching is performed to determine whether it has a true positive or false negative pattern. Pair-wise motif alignment is conducted using SSEARCH between the query sequence and motif sequences of the known family members, to find the best-matched sequence motifs. Once the matched motif regions are confirmed by full-length and motif sequence alignments, they are further modeled with the hidden Markov models against sequence motifs of the respective family. Multiple alignment between the newly identified and the matched motifs are then performed using the ClustalW program (Thompson *et al.*, 1994).

To assist the interpretation of family memberships, overall probability scores for both global and motif matches are provided for top hit families. The global score is computed from the BLAST e-value, the SSEARCH score, and the percentage of sequence identity at overlapped length ratio in SSEARCH alignment. The motif score is computed from the ratio of mismatched amino acids to ProSite patterns, and the hidden Markov motif match score. Family information from ProClass, with hypertext links to all other major family and structural databases, further assists final membership confirmation and the discovery of new relationship. Assisted by the GeneFIND report (Figure 11.4) which contains global similarity, local similarity, and family information all in one place, sequences can be analyzed without laborious human annotation. The system is currently being used by the PIR to assist superfamily classification and database organization.

QueryID: sp|P36136|YK23_YEAST Hypothetical 31.0 KD Protein in GAP1-NAP1 Intergenic Region.
QueryLength: 271aa

A. Most Probable ProSite Patterns: PS00175 PG_MUTASE

GLOBAL	MOTIF	BKID	KLEN	BLAST	SKID	%IDN	RATIO	PSFLAG
2.0	7.0	PMGY_TREPA	251	2e-08	PMGY_TREPA	24.8	0.86	N-PS00175

B. Most Probable PIR Superfamilies: SFA01286 phosphoglycerate mutase

GLOBAL	BKID	KLEN	BSF#	BLAST	SKID	SSF#	%IDN	RATIO
2.0	PMG1_YEAST	246	SFA01286	5e-08	PMG1_YEAST	SFA01286	26.0	0.98

C. Most Probable match to Unclassified Entries:

GLOBAL	BKID	KLEN	BLAST	SKID	%IDN	RATIO
10.0	YK23_YEAST	271	e-152	YK23_YEAST	100.0	1.00

D. ProClass Sequence Entries of Best-Matched Family Members
PCS_AC:SP_ID:SP_AC:PIR_ID:PCF_AC:PS_AC:SFA#:SFA_ALN#:PCM#:LEN:DE
PCS030509:PMG1_YEAST:P00950:PMBYY:PCFA00158Ab:PS00175:SFA01286:-:PCM00175:246|247:Phos..
PCS030533:PMGY_TREPA:P96121:E71357:PCFA00158A#:PS00175:-:-:PCM00175:251|251:Phosphoglycer..
PCS053388:YK23_YEAST:P36136:S38115:PCFC00001:-:-:-:-:271|271: Hypothetical 31.0 KD Protein in GAP..

E. ProClass Family Entries of ProSite Protein Groups and PIR Superfamilies
PCF_AC:PS_AC:PS_DOC#:SFA#:SFA_ALN#:COUNT
PCFA00158Aa:PS00175:PDOC00158:SFA00570:-:11
PCFA00158Ab:PS00175:PDOC00158:SFA01286:-:14
PCFA00158A#:PS00175:PDOC00158:-:-:11
PCFA00158#b:-:-:SFA01286:-:3

F. ProSite Pattern Match and ClustalW Multiple Motif Alignment
AC PS00175
ID PG_MUTASE; PATTERN.
DE Phosphoglycerate mutase family phosphohistidine signature.
PA [LIVM]-x-R-H-G-[EQ]-x(3)-N

```
    PMG1_YEAST+PMBYY      PST      5   LVRHGQSEWN
    PMGY_TREPA+E71357     PST      5   LIRHGESEWN
    QueryID               GFN     10   IVRHGQTEWS
                                       ::***::**.
```

G. SSEARCH Full-Length Sequence Alignment to Best-Matched Family Members (not shown)

Figure 11.4 GeneFIND report: an example.

11.5 References

Altschul, S. F., Madden, T. L., Schaffer, A. A., Zhang, J., Zhang, Z., Miller, W. & Lipman, D. J. (1997) Gapped BLAST and PSI-BLAST: a new generation of protein database search programs. *Nucleic Acids Res* **25**, 3389-402.

Blom, N., Hansen, J., Blaas, D. & Brunak, S. (1996). Cleavage site analysis in picornaviral polyproteins: discovering cellular targets by neural networks. *Protein Sci* **5**, 2203-16.

Claros, M. G., Brunak, S. & von Heijne, G. (1997). Prediction of N-terminal protein sorting signals. *Curr Opin Struct Biol* **7**, 394-8.

Elhammer, A.P., Poorman, R,A,, Brown, E., Maggiora, L.L., Hoogerheide, J.G. & Kezdy, F. J. (1993). The specificity of UDP-GalNAc:polypeptide N-acetylgalactosaminyltransferase as inferred from a database of *in vivo* substrates and from the in vitro glycosylation of proteins and peptides. *J Biol Chem* **268**, 10029-38.

Ferran, E. A. & Ferrara, P. (1992). Clustering proteins into families using artificial neural networks. *Comput Appl Biosci* **8**, 39-44.

Ferran, E. A. & Pflugfelder, B. (1993). A hybrid method to cluster protein sequences based on statistics and artificial neural networks. *Comput Appl Biosci* **9**, 671-80.

Ferran, E. A., Pflugfelder, B. & Ferrara, P. (1994). Self-organized neural maps of human protein sequences. *Protein Sci* **3**, 507-21.

Gierasch, L. M. (1989). Signal sequences. *Biochemistry* **28**, 923-30.

Goldberg, D. E. (1989). *Genetic Algorithms in Search, Optimization and Machine Learning*. Addison-Wesley, Redwood City, CA.

Gulukota, K., Sidney, J., Sette, A. & DeLisi, C. (1997). Two complementary methods for predicting peptides binding major histocompatibility complex molecules. *J Mol Biol* **267**, 1258-67.

Hansen, J. E., Lund, O., Engelbrecht, J., Bohr, H., Nielsen, J. O. & Hansen, J. E. (1995). Prediction of O-glycosylation of mammalian proteins: Specificity patterns of UDP-GalNAc:polypeptide N-acetylgalactosaminyl transferase. *Biochem J* **308**, 801-13.

Hansen, J. E., Lund, O., Tolstrup, N., Gooley, A. A., Williams, K. L. & Brunak, S. (1998). NetOglyc: prediction of mucin type O-glycosylation sites based on sequence context and surface accessibility. *Glycoconj J* **15**, 115-30.

Holland, J. H. (1992). *Adaptation in Neural and Artificial Systems*. MIT P, Cambridge.

Koza, J. R. (1992). *Genetic Programming: On the Programming of Computers by Means of Nature Selection*. MIT Press, Cambridge, MA.

Ladunga, I., Czako, F., Csabai, I. & Geszti, T. (1991). Improving signal peptide prediction accuracy by simulated neural network. *Comput Appl Biosci* **7**, 485-7.

Lohmann, R., Schneider, G., Behrens, D. & Wrede, P. (1994). A neural network model for the prediction of membrane-spanning amino acid sequences. *Protein Sci* **3**, 1597-601.

Lohmann, R., Schneider, G. & Wrede, P. (1996). Structure optimization of an artificial neural filter detecting membrane-spanning amino acid sequences. *Biopolymers* **38**, 13-29.

Mezard, M. & Nadal, J. P. (1989). Learning in feedforward layered networks: The Tilting algorithm. *J Phys A* **22**, 2193-203.

Nakata, K. (1995). Prediction of zinc finger DNA binding protein. *Comput Appl Biosci* **11**, 125-31.

Nielsen, H., Engelbrecht, J., von Heijne, G. & Brunak, S. (1996). Defining a similarity threshold for a functional protein sequence pattern: the signal peptide cleavage site. *Proteins* **24**, 165-77. (The data set is available from the FTP server at ftp://virus.cbs.dtu.dk/pub/signalp).

Nielsen, H., Engelbrecht, J., Brunak, S. & von Heijne, G. (1997). Identification of prokaryotic and eukaryotic signal peptides and prediction of their cleavage sites. *Protein Eng* **10**, 1-6.

Pearson, W. R. (1991). Searching protein sequence libraries: comparison of the sensitivity and the selectivity of the Smith-Waterman and FASTA algorithms. *Genomics* **11**, 635-50.

Pisano, A., Redmond, J.W., Williams, K.L. & Gooley, A. A. (1993). Glycosylation sites identified by solid-phase Edman degradation: O-linked glycosylation motifs on human glycophorin A. *Glycobiology* **3**, 429-35.

Rapoport, T. A. (1992). Transport of proteins across the endoplasmic reticulum membrane. *Science* **258**, 931-6.

Rooman, M. J., Kocher, J. P. & Wodak, S. J. (1992). Extracting information on folding from the amino acid sequence: accurate predictions for protein regions with preferred conformation in the absence of tertiary interactions. *Biochemistry* **31**, 10226-38.

Schneider, G., Schuchhardt, J. & Wrede, P. (1994). Artificial neural networks and simulated molecular evolution are potential tools for sequence-oriented protein design. *Comput Appl Biosci* **10**, 635-45.

Schneider, G., Schuchhardt, J. & Wrede, P. (1995). Peptide design in machina: development of artificial mitochondrial protein precursor cleavage sites by simulated molecular evolution. *Biophys J* **68**, 434-47.

Smith, T. F. & Waterman, M. S. (1981) Comparison of bio-sequences. *Adv Appl Math* **2**, 482-9.

Thompson, J.D., Higgins, D.G. & Gibson, T.J. (1994) CLUSTAL W: improving the sensitivity of progressive multiple sequence alignment through sequence weighting, position specific gap penalties and weight matrix choice. *Nucleic Acids Res* **22**, 4673-80.

von Heijne, G. (1983). Patterns of amino acids near signal-sequence cleavage sites. *Eur J Biochem* **133**, 17-21.

von Heijne, G. (1985). Signal sequences. The limits of variation. *J Mol Biol* **184**, 99-105.

von Heijne, G. (1986). A new method for predicting signal sequence cleavage sites. *Nucleic Acids Res* **14**, 4683-90.

von Heijne, G. & Abrahmsen, L. (1989). Species-specific variation in signal peptide design. Implications for protein secretion in foreign hosts. *FEBS Lett* **244**, 439-46.

von Heijne, G. (1995). Protein sorting signals: simple peptides with complex functions. *EXS* **73**, 67-76

Wilson, I. B., Gavel, Y. & von Heijne, G. (1991). Amino acid distributions around O-linked glycosylation sites. *Biochem J* **275** (Pt 2), 529-34.

Wrede, P., Landt, O., Klages, S., Fatemi, A., Hahn, U. & Schneider, G. (1998). Peptide design aided by neural networks: biological activity of artificial signal peptidase I cleavage sites. *Biochemistry* **37**, 3588-93.

Wu, C., Whitson, G., McLarty, J., Ermongkonchai, A. & Chang, T. C. (1992). Protein classification artificial neural system. *Protein Sci* **1**, 667-77.

Wu, C. H., Berry, M., Shivakumar, S. & McLarty, J. (1995). Neural networks for full-scale protein sequence classification: Sequence encoding with singular value decomposition. *Machine Learning* **21**, 177-93.

Wu, C. H. (1996). Gene classification artificial neural system. *Methods Enzymol* **266**, 71-88.

Wu, C. H., Zhao, S., Chen, H. L., Lo, C. J. & McLarty, J. (1996). Motif identification neural design for rapid and sensitive protein family search. *Comput Appl Biosci* **12**, 109-18.

Wu, C. H., Huang, H. & McLarty, J. (1999a). Gene family identification network design. *International Journal on Artificial Intelligence Tools*, special Biocomputing issue (in press).

Wu, C. H., Shivakumar, S. & Huang, H. (1999b). ProClass Protein Family Database. *Nucleic Acids Res* **27**, 272-4.

Part IV

Open Problems and Future Directions

It has been shown in this book that neural networks have been applied extensively to problems in genome informatics. The success of these applications ensures that the field will likely continue in similar directions, at least in the near future. However, this is not to say that neural networks should be used for genome informatics problems to the exclusion of other, more traditional approaches. There are many ways to approach problems of modeling, classification and prediction: the goal should be, for a given situation, to use the best technique applicable, whether that technique be statistical analysis or neural networks or a combination of techniques. Problems inhere in both statistical and neural network approaches. Hybrid neural network techniques have been developed to solve some of the problems with traditional neural networks. There are numerous kinds of artificial neural network architectures, only a few of which have been applied to genome informatics. Choices of which technology to use and which combinations can be used effectively are open problems. There are also complementary fields of research, such as *fuzzy logic* and *genetic algorithms*, that may be combined with neural networks to effect solutions that neither technology can do alone.

Major problems facing an investigator who wants to prepare data for analysis or neural network modeling concern what input data features are to be used and how the information will be encoded before presentation to the model. Another issue to be faced concerns discovery of biological rules and features from the data, after analysis or modeling--e.g., what do the results mean? Interpretation of weights after training, for example, is a particularly difficult problem.

CHAPTER 12

Integration of Statistical Methods into Neural Network Applications

Numerous publications have addressed the issue of similarity between neural network techniques and traditional statistical methods (Cheng & Titterington, 1994; Sarle, 1994). Clearly, the two approaches overlap somewhat and, in some situations, are even identical. For example, ordinary linear regression can be achieved with a simple perceptron. Linear discriminant analysis is equivalent to classification by a perceptron. A neural network with no hidden units and a single logistic output unit is functionally equivalent to logistic regression, a statistical tool widely used for analysis of binary data. Kohonen self-organizing maps are similar to kernel regressions and k-means clustering (Cheng, 1994). Many other examples could be given. The question has been asked of those who apply neural network technology to genome informatics research, why use neural networks and not standard statistical methods? This is not the right question. The question should be, how can both schools of thought be better used together to address the fascinating but often very difficult problems in genome informatics? Neural networks should be used because they work so well, and the same can be said for standard statistical methods. Many successful examples of neural network applications have been presented in this book, but there are many similarly successful examples of statistical applications not included in this book. Neural networks provide a good framework for the presentation of familiar statistical concepts, and some statistical techniques can be implemented using neural network technology (Cheng & Titterington, 1994). Some neural networks even have probabilistic elements, and Bayesian techniques are increasingly being restated into neural network terminology. Perhaps the biggest appeal of neural networks to researchers in genome informatics is the highly visual approach to problem solving and the (presumed) relaxation of the formal restrictions required by statistical approaches. It is easy to conceptualize solutions to difficult highly nonlinear problems with the visual neural network approach. Multilayer perceptrons are good for nonlinear data fitting and prediction and nonlinear problems abound in informatics. However, it must be recognized that neural networks are not always preferable to standard statistical techniques (Duh *et al.*, 1998). Statistical tools need to be integrated into the use of neural network tools. Many important problems in data preparation, model building and evaluation of results could benefit from statistical methodology and thinking. As an example: in statistical regression, the significance of a regression coefficient can be estimated and tested to see if it is significantly different from 0, i.e., whether it should be included in the regression equation or not. This is equivalent to testing whether one of the input or hidden units in a neural network is needed. There is no widely used equivalent formal method for determining the value of a unit in a neural network, but such methods are in development (to be discussed below). Similarly, rigorous methods for testing the value of one neural network architecture versus another need to be developed. Estimation of confidence limits of neural network output results could greatly improve the usefulness of applications. Nonlinear regression has a vast literature and so has the field of numerical function minimization. However, many of the techniques in those fields were either reinvented by neural network developers or remain yet to be applied to neural networks. Historically, the problems mentioned above have often been addressed heuristically or ignored in the neural network literature. One reason for this neglect is that the statistical issues can be quite complex. Another reason for neglect of the statistical aspects of neural networks is that the people applying neural networks are usually not statisticians. The successful marriage of neural network technology and statistical methods remains one of the important open problems for future research. There is no need for a neural network/statistics dichotomy. There is a need for both fields to be used jointly to better solve interesting problems.

12.1 Problems in Model Development

Although the visual and highly intuitive methods of artificial neural networks make the process of nonlinear modeling much more available to non-statisticians, the problems encountered are not effectively different from those encountered in using statistical modeling methods. The number and relationship of variables (input units) to be used as input to the model must still be determined, the number of parameters (hidden layers, hidden units) remains to be determined and methods for evaluating the results (output) applied. If various models are possible, there must be means of selecting among them.

12.1.1 Input Variable Selection

The problem of preparing data for input to neural networks has been, in part, discussed previously as methods of *feature selection* and *data encoding*. The essence of the problem is to prepare input vectors for training that contain all the information necessary to train a network for its particular task. Statistical regression usually adheres to rigorous statistical methods for determining the value of a particular input variable or a set of variables, but even in regression the answers are not necessarily unique. For example, forward stepwise procedures (adding the most significant variables to the regression, one at a time, and checking at each step to see if any previously entered variables could be removed) may yield a different set of variables than backward stepwise procedures (starting with all variables in the regression and then checking in a stepwise fashion to see if any could be removed). This procedure can be even more complex for nonlinear regression. Such regression techniques yield a number (usually a probability or a likelihood or a confidence interval) that leads to a decision whether a given variable or set of variables is essential to the regression. Few attempts have been made at systematic input variable selection for neural networks, at least as applied to genome informatics. Bootstrapping techniques have been applied in a clinical setting for input variable selection (Baxt & White, 1995), and second-order derivatives have been used for the same purpose in a technique unfortunately called *optimal brain damage* (Hassibi & Stork, 1993; Stahlberger & Riedmiller, 1997). Optimal brain damage is a method of pruning or eliminating weights based on the magnitude of the second derivatives of the error function with respect to the weights: unimportant weights are eliminated, the network retrained and pruning continued until no more can be eliminated. This is a reasonably successful method, but it can be quite computationally intensive. Linear approximation methods and orthogonal least squares methods have been successfully used for input variable selection in some applications--in particular, for predicting properties of molecules from their structure (Duprat *et al.*, 1998). The choice of input variables for neural networks is an important area that needs considerable work by both statisticians and neural network scientists. The problem of input data encoding is especially difficult with genome data. Neural networks must have, by their nature, a fixed number of input units; genome data, on the other hand, requires dealing with records of variable length (e.g. proteins in a family are of differing lengths). Some methods for handling this problem are illustrated in Part III of this book. Encoding genome data for neural networks still remains one of the most difficult tasks.

12.1.2 Number of Hidden Layers and Units

Theoretically, in a multilayer perceptron, no more than two hidden layers should be required for any task. For most applications, one layer is sufficient. But there is no way to determine in advance how many layers are preferred for a given problem. For a given number of input layers, the same problem exists for the number of hidden units: how many are needed? Typical applications have from one to hundreds (or even thousands) of hidden units. Again, what is needed is a systematic, statistical, means of discarding unnecessary hidden units. Pruning by looking at second-order derivatives can also work for this problem, but clearly more work is needed. Other methods include working with likelihood functions and Bayes ratios (Ripley, 1997b). Koene and Takane (1999) use a method called *discriminant components* pruning to reduce the complexity of large networks. Cotrell and others have developed a stepwise method for weight elimination (Cottrell *et al.*, 1995).

12.1.3 Comparison of Architectures

A more global problem concerns the comparison of different architectures, or even techniques (e.g. nonlinear regression versus neural networks), as applied to the same problem. Which solution is the best for a given application? A statistical way to embed several such solutions into a related family of techniques is needed, with a rigorous, automatic way of determining which solution best fits a given problem. There have been several approaches to this problem. Complications arise in part from the lack of a unique global minimum in the error function--back-propagation can yield many equivalent solutions. One approach has been to compare the error function results obtained with different models applied to the same data, using receiver operating characteristics (ROC) methodology to evaluate the results (Metz *et al.*, 1998a; Metz *et al.*, 1998b). An ROC curve is a plot of sensitivity versus $1 -$ specificity; better fits are indicated by curves that have a larger area under the curve and that approach an ideal curve. More work is needed to predict which architecture or technique will perform best for a given application and to estimate the magnitude of the difference between techniques.

12.1.4 Need for Benchmark Data

One problem with a rapidly growing field of research is the lack of standardized sets of data for training and evaluation, i.e., *benchmark data*. Whenever developers use different sets of data for publications and presentations, comparing the efficacy of new methods becomes extremely difficult. If benchmark data sets were readily available and widely used, then comparison of methods would be more meaningful. At least one international conference (Neural Information Processing Systems – NIPS 95) and several web sites have been devoted to this task. For example, the University of Toronto has a web site with benchmark data for neural network applications, called DELVE. Proben1(Prechelt, 1994) is another WWW-based set of benchmarks and benchmarking rules for neural network training algorithms. However, there are few such benchmarks available for genome informatics. Some were discussed in Chapter 8 and others exist (Bailey *et al.*, 1998; Bohm, 1995; Bohm, 1997), but more work is needed in this area.

12.2 Training Issues

In spite of the popularity of back-propagation and its use for decades, serious problems still exist with training neural networks. Back-propagation can be very slow to converge; solutions are not unique and may not represent a global minimum; and stopping criteria are, for the most part, heuristic rather than theoretically based. Bayesian methods have been proposed to eliminate the need for validation data (Ripley, 1997a), but are not widely used. Many variants of back-propagation have been developed to speed up convergence.

Yet another problem, one more appropriately addressed by researchers in genome informatics, is the selection of training data. Obviously, training data should be representative of the data to which a network will be applied--the test data. Random sampling is one method for selecting representative data. However, researchers should be more aware of how training data selection will influence the performance of the neural network.

12.3 Interpretation of Results

A major criticism of neural networks is that it is difficult to interpret the meaning of the weights, after training. A network's performance may be acceptable but lend no insight into how it works, a problem reminiscent of principal component analysis: component functions are often hard to interpret. This is a difficult and important challenge for future research.

Confidence in results is another problem related to statistical interpretation. Ordinary regression techniques can yield a prediction with confidence limits around the prediction to aid in the use and interpretation of results. Bootstrapping (Lajbcygier & Connor, 1997) and other techniques (Heskes, 1997; Hwang & Ding, 1997) have been developed to estimate confidence limits for neural networks but are not widely applied. Bayesian techniques and probabilistic networks (Ripley, 1993; Specht, 1990) also help with this problem: the outputs can be interpreted as probabilities. When there are more than two output categories--in classification, for example--a trick called a *softmax* function (Bridle, 1990a; Bridle, 1990b) (actually a multiple logistic function) has been used to restrict the total of the outputs to 1.0, in effect making them interpretable as probabilities.

12.4 Further Sources of Information

For a discussion of statistical methods and neural networks, see the book by Ripley (1996) and articles by Ripley (1993), Cheng (Cheng & Titterington, 1994) and Sarle (1994). Warren Sarle maintains an excellent neural network FAQ (frequently asked questions) web page for the Usenet newsgroup comp.ai.neural-nets; this web page contains many tutorial discussions and references to books, reviews, journal articles and other neural network resources.

12.5 References

Bailey, L. C., Jr., Searls, D. B. & Overton, G. C. (1998). Analysis of EST-driven gene annotation in human genomic sequence. *Genome Res* **8,** 362-76.

Baxt, W. G. & White, H. (1995). Bootstrapping confidence intervals for clinical input variable effects in a network trained to identify the presence of acute myocardial infarction. *Neural computation* **7,** 624-638.

Bohm, K. (1995). High performance computing for the human genome project. *Comput Methods Programs Biomed* **46,** 107-12.

Bohm, K. (1997). Supercomputing in cancer research. *Stud Health Technol Inform* **43 Pt A,** 104-8.

Bridle, J. S. (1990a). Probabilistic interpretation of feedforward classification network outputs, with relationships to statistical pattern recognition. In *Neurocomputing: Algorithms, Architectures and Applications* (ed. F. F. Soulie and J. Herault), pp. 227-36. springer-Verlag, Berlin.

Bridle, J. S. (1990b). Training stochastic model recognition algorithms as networks can lead to maximum mutual information estimation of parameters. In *Advances in Neural Information Processing Systems*, vol. 2 (ed. D. S. Touretzky), pp. 211-17. Morgan Kaufmann, San Mateo.

Cheng, B. & Titterington, D. M. (1994). Neural Networks: a review from a statistical perspective. *Statistical Sciences* **9,** 2-54.

Cottrell, M., Girard, B., Girard, Y., Mangeas, M. & Muller, C. (1995). Neural modeling for time series: a statistical stepwise method for weight elimination. *IEEE Transactions on Neural Networks* **6,** 1355-1364.

Duh, M. S., Walker, A. M., Pagano, M. & Kronlund, K. (1998). Prediction and cross-validation of neural networks versus logistic regression: using hepatic disorders as an example. *Am J Epidemiol* **147,** 407-13.

Duprat, A. F., Huynh, T. & Dreyfus, G. (1998). Toward a principled methodology for neural network design and performance evaluation in QSAR. Application to the prediction of logP. *J. Chem. Inf. Comput Sci.* **38,** 586-94.

Hassibi, B. & Stork, D. G. (1993). Second order derivatives for network pruning: Optimal Brain Surgeon. In *Advances in Neural Information Processing Systems*, vol. 5 (ed. S. J. Hanson, J. D. Cowan and C. L. Giles), pp. 263-270. Morgan-Kaufmann, San Mateo, Ca.

Heskes, T. (1997). Practical confidence and prediction intervals. In *Advances in Neural Information Processing Systems*, vol. 9 (ed. M. C. Mozer, M. I. Jordan and T. Petsche), pp. 176-82. The MIT Press, Cambridge, MA.

Hwang, J. T. G. & Ding, A. A. (1997). Prediction intervals for artificial neural networks. *American Statistical Association* **92,** 748-57.

Koene, R. A. & Takane, Y. (1999). Discriminant component pruning. Regularization and interpretation of multi-layered back-propagation networks. *Neural Comput* **11,** 783-802.

Lajbcygier, P. R. & Connor, J. T. (1997). Improved option pricing using artificial neural networks and bootstrap methods. *Int J Neural Syst* **8,** 457-71.

Metz, C. E., Herman, B. A. & Roe, C. A. (1998a). Statistical comparison of two ROC-curve estimates obtained from partially-paired datasets. *Med Decis Making* **18,** 110-21.

Metz, C. E., Herman, B. A. & Shen, J. H. (1998b). Maximum likelihood estimation of receiver operating characteristic (ROC) curves from continuously-distributed data. *Stat Med* **17,** 1033-53.

Prechelt, L. (1994). PROBEN1 - A set of benchmarks and benchmarking rules for neural network training algorithms. Fakultat fur Informatik, Universitat Karlsruhe, Karlsruhe, Germany.

Ripley, B. D. (1993). Statistical aspects of neural networks. In *Networks and Chaos - Statistical and Probabilistic Aspects* (ed. O. E. Barndorff-Nielsen, J. L. Jensen and W. S. Kendall), pp. 40-123. Chapman & Hall, London.

Ripley, B. D. (1996). *Pattern recognition and neural networks*. Cambridge University Press.

Ripley, B. D. (1997a). Can statistical theory help us use neural networks better? In *Interface 97. 29th Symposium on the Interface: Computing Science and Statistics*.

Ripley, B. D. (1997b). Statistical ideas for selecting network architectures. In *Neural Networks: Artificial Intelligence and Industrial Applications* (ed. K. B. and S. Gielen), pp. 183-90. Springer.

Sarle, W. S. (1994). Neural networks and statistical models. In *Nineteenth Annual SAS Users Group International Conference*, pp. 1538-50, Cary, NC: SAS Institute.

Specht, D. F. (1990). Probabilistic neural networks. *Neural Networks* **2,** 110-18.

Stahlberger, A. & Riedmiller, M. (1997). Fast network pruning and feature extraction using the Unit-OBS algorithm. In *Advances in Neural Information Procession Systems*, vol. 9 (ed. M. C. Mozer, M. I. Jordan and T. Petsche), pp. 655-661. the MIT Press, Cambridge, MA.

CHAPTER 13

Future of Genome Informatics Applications

Several areas hold promise for continuing research. One attractive feature of the neural network method is that it requires no *a priori* knowledge of the application problem in order to build the model. Nevertheless, to make the neural network technology truly useful, it is highly desirable to be able to extract knowledge from the trained networks and discover the underlying biological rules (13.1). Meanwhile, it is equally important that we apply our existing knowledge, if any, to improve the neural network design (13.2). We will conclude this chapter, and the book, with an analysis of the genome informatics applications regarding the capacities of the neural networks and the rationales behind their choice, and contemplate the future of neural networks for biology (13.3).

13.1 Rule and Feature Extraction from Neural Networks

Neural networks are often viewed as black boxes. Despite the high level of predictive accuracy, one usually cannot understand why a particular outcome is predicted. Although this is generally true, especially for multilayer networks whose weights can not be easily interpreted, there are methods for analyzing trained networks and extracting rules or features. The issue can be framed as different set of questions. How does one extract rules from trained networks (13.2.1)? Is it possible to measure the importance of inputs (13.2.2)? How should input variables be selected (13.2.3)? Another related question concerns the interpretation of network output: How likely is the prediction to be correct (13.2.4)?

13.2.1 Rule Extraction from Pruned Networks

The process of extracting rules from a trained network can be made much easier if the complexity of the network has first been reduced. Furthermore, it is expected that fewer connections will result in more concise rules. Setiono (1997a) described an algorithm for extracting rules from a pruned network. The network was a standard three-layer feedforward back-propagation network trained with a pre-specified accuracy rate. The pruning process attempted to eliminate as many connections as possible while maintaining the accuracy rate.

One of the examples used to illustrate the rule extraction algorithm was the intron-exon splice junction problem where donor and acceptor sites were to be distinguished from other sequences containing no splice junction. The original neural network had 241 input units (240 units for BIN4 encoding of sequence windows of 60 bases long, plus a bias unit), 5 hidden units, and 3 output units (for the three classes of donor site, acceptor site and no site). The pruning algorithm applied a penalty function for weight elimination (Setiono, 1997b). The method required no initial knowledge of the problem domain; instead, relevant and irrelevant attributes of the data were distinguished during the pruning process. When redundant weights were eliminated, redundant input and hidden units were identified and removed from the network.

The resulting pruned network had only 16 weights remaining, significantly reduced from the original 1220 (241 x 5 + 5 x 3) weights. Of the original 241 inputs, only 10 input units (including the bias input unit) remained; and only 3 out of the 5 hidden units were retained. Subsequently, ten rules were extracted from the pruned network and applied to splice junction prediction with the same accuracy rate of the pruned network. In this particular study, the extracted rules were 7-base-long patterns, which resembled the consensus patterns generally described for splice junctions (such as the AG and GT pairs). It is not clear, however, whether at least some of the rules represent patterns that cannot be directly derived from the training data. This would indicate whether the rules (i.e., network-derived patterns) did capture input features (e.g., co-variation of bases) that would otherwise be unavailable from the consensus method, which assumes positional independence. It would also be interesting to perform a direct comparison of the predictive accuracies of the original three-layer network, the pruned network and its extracted rules, a two-layer perceptron, as well as a weight matrix and consensus patterns derived from the training data. Although the rule extraction algorithm presented above cannot be fully assessed without a thorough comparative study, the idea does represent an important future research direction for knowledge extraction from trained neural networks.

13.2.2 Feature Extraction by Measuring Importance of Inputs

Various methods have been proposed to measure the importance of inputs (Sarle, 1998) and are likely to be useful in different applications of neural nets. The two most common notions of *importance* are *predictive importance* and *causal importance*. Predictive importance is concerned with the increase in generalization error when an input is omitted from a network. Causal importance is concerned with situations in which an individual wants to quantify the relationship between input value manipulation and consequent output change.

In linear models, the weights have a simple interpretation: each weight is the change in the output associated with a unit change in the corresponding input, assuming all other inputs are held fixed. Thus, if all the inputs can change independently of each other, then the weights are directly related to *causal importance* of the inputs. Due to the hidden layer(s), the weights in a multilayer neural network cannot be easily interpreted like those in a linear model. Nevertheless, the gradient of the output (i.e., a vector of partial derivatives) with respect to the inputs has the same interpretation as weights in a linear model. Since the partial derivative of the output with respect to each input provides only local information, it might be better to look at the change in the output over an interval to measure the causal importance.

Another way to measure the importance of an input is to omit it from the model, retrain the model, and see how much the error function increases. The change in the error function is a direct measure of *predictive importance*, but this measure can be misleading when the inputs are correlated. Since retraining the network can be time-consuming if there are many more training cases than weights in the network, using the Hessian matrix may be more efficient to approximate the change in the error function. This method was illustrated in a network pruning algorithm, called *optimal brain surgeon* (Stahlberger & Riedmiller, 1997; Hassibi & Stork, 1993). If the inputs are statistically dependent, measuring the importance of inputs becomes even more difficult because the effects of different inputs cannot generally be separated.

A recent example of feature extraction using the predictive importance is seen in Jones (1999). From the simple network architecture of 6 (input) x 6 (hidden) x 2 (output units), the relative importance of the input features for protein fold recognition, such as the sequence profile score and the pairwise energy term, was analyzed. Likewise, Nair (1997) analyzed the *E. coli* ribosome binding sites by selectively changing sequences with randomized calliper inputs and observing the corresponding neural network performance. The results showed that the initiation codon and the Shine-Dalgarno sequence already known to be important for translation initiation were also important in imparting knowledge to the network.

13.2.3 Feature Extraction Based on Variable Selection

One of the most common problems encountered in data modeling is the choice of independent variables for inclusion in the model. Perhaps one of the major disadvantages of any variable selection method based on an underlying linear model is the fact that the form of the model is specified in advance. If some nonlinear model better fits the data, then this may be used as the underlying model for the variable selection process. A neural network properly fitted to a data set should make use of the best nonlinear model as dictated by the data itself. Thus a variable selection procedure applied to the network model might be expected to extract the most relevant set of variables, at least in terms of modeling the output (response) variable.

Network pruning techniques can also be considered as variable selection methods, since they generally aim to prune unnecessary connections and neurons. Several pruning algorithms were studied (Livingstone *et al.*, 1997), two based on the direct analysis of neuron weights (*magnitude-based*), and three others on change in network error following the elimination of connections (*error-based*). The utility of the network pruning method for variable selection was then illustrated by applying one of the magnitude-based methods to a chemical system involving the formation of charge-transfer complexes between monosubstituted benzenes and a common electron acceptor, 1,3,5-trinitrobenzene. Eleven physicochemical parameters were selected, based on their individual correlations with the charge-transfer substituent constant (k), from a set of 58 parameters. The three-layer feedforward back-propagation network had 11 input units for the 11 physicochemical properties, 5 hidden units, and a single output unit for k. The result of the variable selection showed that some of the variables with higher correlations were eliminated before others with lower correlations. This indicated some underlying nonlinear relationships between k and the physicochemical descriptors. Presumably, it is this nonlinearity which was responsible for the different order of variable selection by the network, although the colinearity or multicollinearity amongst the descriptors might also contribute.

13.2.4 Network Understanding Based on Output Interpretation

Another aspect of understanding the meaning of a trained network and assessing the significance of its prediction is the interpretation of network output. This is especially important when there is limited knowledge in the application domain not only regarding what constitutes important input features, but also concerning low likelihood a given prediction is correct.

One approach for assigning a natural meaning to the output scores is to provide a probabilistic point of view. In such a framework, the output scores are combined as a linear weighted sum, and the sum and scores are interpreted as log probabilities. A *softmax* output activation function is often used to make the sum of the outputs equal to one, so that the outputs are interpretable as posterior probabilities (Bridle, 1990). The softmax function, described previously, reduces to the simple logistic function when there are only two output categories. Therefore, when there are only two categories, as seen in the sequence discrimination problem with a yes/no answer, it is simpler to use just one output unit with a logistic output activation function. The probabilistic interpretation allows one to assess the confidence of network prediction, which in turn assists the extraction of significant input features. Riis and Krogh (1996) used the softmax function in their secondary structure prediction networks, so that the network outputs could be interpreted as the estimated probabilities of correct prediction to indicate which residues are predicted with high confidence.

Neural networks, if properly trained, can generalize and predict well for unseen, related data. But what if the task involves classifying data dissimilar to any of the training cases? To detect novel cases for which the classification may not be reliable, one can apply a classification method based on probability density estimation, such as a *probabilistic neural network* (Specht, 1990). The network is especially useful in revealing whether a test case is similar (i.e. has a high density) to any of the training data if all inputs are relevant. Probabilistic neural network can be considered as a normalized *radial basis function network* (Chapter 4.1) in which there is a hidden unit centered at every training case. These radial basis function units are called *kernels* and are usually probability density functions such as the *Gaussian*.

13.2 Neural Network Design Using Prior Knowledge

Current trends in neural networks favor smaller networks with minimal architecture. Two major advantages of smaller networks previously discussed are better generalization capability (8.4) and easier rule extraction (13.2). Another advantage is better predictive accuracy, seen when a large network is replaced by many smaller networks, each for a subtask or a subset of data. A typical example is the protein classification problem, where n individual networks can be used to classify n different protein families and increase the prediction accuracy obtained by one large network with n output units. The improvement is especially significant when there is sufficient data for fine-tuning individual neural networks to the particularity of the data subsets. The use of *ensembles* of small, *customized* neural networks to improve predictive accuracy has been shown in numerous cases.

The customization may involve using different encoding methods or network architectures to extract characteristic features (i.e., *a priori* knowledge) of the data. Instead of using the same generic network model to predict the three types of secondary structures (α-helix, β-sheet, and coil), Riis and Krogh (1996) designed a helix-network with a built-in period of three residues in its connections to capture the characteristic periodic structure of helices. Wu *et al.* (1996) used a term weighting method in sequence encoding and additional feature units to emphasize n-gram terms conserved in different protein families. Dubchak *et al.* (1997) observed the effects of specific amino acid properties on different tertiary structural (folding) classes, which can be used to develop individual sets of descriptors. Nielsen *et al.* (1997) trained specific networks with optimized architectures (e.g., the sequence window length, hence the input vector size) to account for the wide length variations of signal peptides from different taxonomic origins, such as eukaryotic organisms, and gram-positive and gram-negative bacteria.

Several other recent studies assert the importance of incorporating existing knowledge into the encoding methods (Chapter 7). For example, Reese *et al.* (1997) showed that the change of encoding method alone, to account for the pairwise correlations between adjacent nucleotides, resulted in a 5% increase in gene structure predictive accuracy (Chapter 9.2).

13.3 Conclusions

We can sum up what one can do with a neural network. In principle, neural networks are universal approximators and can compute any computable function. In practice, neural networks are especially useful for classification and function approximation/mapping problems that have plenty of training data available and can tolerate some imprecision but that resist the easy application of hard and fast rules.

How have neural networks been used in genome informatics applications? In Part II, we have summarized them based on the types of applications for DNA sequence analysis, protein structure prediction and protein sequence analysis. Indeed, the development of neural network applications over the years has resulted in many successful and widely used systems. Current state-of-the-art systems include those for gene recognition, secondary structure prediction, protein classification, signal peptide recognition, and peptide design, to name just a few.

We can also analyze the types of neural networks that have been developed for genome informatics and the reasons for the choices. One of the most common uses of neural networks for sequence analysis problems has been *sequence discrimination*, i.e., for differentiating positive from negative cases. The site detection problem (i.e., the recognition of DNA splice sites and various binding or functional sites on DNA or protein sequences) is commonly cast as a sequence discrimination problem. What are the biological rules involved in the recognition of such sites and how are they captured by different methods? The common approaches of weight matrix or consensus pattern searching are based on the observations of *positive* sequence features associated with these sites, and assume *independence* among individual residues surrounding the sites. Known biological rules, however, indicate that site recognition by macromolecules may involve both *positive and negative* features, and that neighboring or adjacent residues are often *correlated*. Neural networks are trained from both positive and negative examples, and can account for the correlation among residues in their neural connections. Thus, they have become a popular choice for sequence discrimination tasks.

Neural network method is often quoted as a *data-driven* method. The weights are adjusted on the basis of data. In other words, neural networks learn from training examples and can generalize beyond the training data. Therefore, neural networks are often applied to domains where one has little or incomplete understanding of the problem to be solved, but where training data is readily available. Protein secondary structure prediction is one such example. Numerous rules and statistics have been accumulated for protein secondary structure prediction over the last two decades. Nevertheless, these features are highly *interdependent* and *nonlinear* in nature, and would be nontrivial to fit. Several other genome informatics problems also involve multiple features, such as the various exon-coding potentials and fold recognition scores. As a *data modeling* method, neural networks can combine the multiple features or rules, and fit complex nonlinear models. Furthermore, unlike other nonlinear statistical modeling techniques, these models do not have to be specified in advance; thus, one can avoid making assumptions that may not be correct or relevant for the biological domains.

The many features and advantages of the neural network method have made it an important research tool for genome informatics. As a useful adjunct to other statistical and mathematical methods, neural networks will continue to play important roles in life science, where complex biological knowledge cannot be easily modeled, and will help us understand and answer fundamental biological questions.

13.4 References

Bridle, J. S. (1990). Probabilistic interpretation of feedforward classification network outputs, with relationships to statistical pattern recognition. In: *Neurocomputing: Algorithms, Architectures and Applications*, (Fogleman Soulie, F. & Herault, J., eds.), Springer-Verlag, Berlin. pp. 227-36.

Dubchak, I., Muchnik, I. & Kim, S. H. (1997). Protein folding class predictor for SCOP: approach based on global descriptors. *Ismb* **5,** 104-7

Hassibi, B. & Stork, D.G. (1993). Second order derivatives for network pruning: Optimal Brain Surgeon. *Advances in Neural Information Processing Systems* **5,** 164-71.

Jones, D. T. (1999). GenTHREADER: An efficient and reliable protein fold recognition method for genomic sequences. *J Mol Biol* **287,** 797-815.

Livingstone, D. J., Manallack, D. T. & Tetko, I. V. (1997). Data modelling with neural networks: advantages and limitations. *J Comput Aided Mol Des* **11,** 135-42.

Nair, T. M. (1997). Calliper randomization: an artificial neural network based analysis of *E. coli* ribosome binding sites. *J Biomol Struct Dyn* **15,** 611-7.

Nielsen, H., Engelbrecht, J., Brunak, S. & von Heijne, G. (1997). Identification of prokaryotic and eukaryotic signal peptides and prediction of their cleavage sites. *Protein Eng* **10,** 1-6.

Reese, M. G., Eeckman, F. H., Kulp, D. & Haussler, D. (1997). Improved splice site detection in Genie. *J Comput Biol* **4,** 311-23.

Riis, S. K. & Krogh, A. (1996). Improving prediction of protein secondary structure using structured neural networks and multiple sequence alignments. *J Comput Biol* **3,** 163-83.

Sarle, W. S. (1998). How to measure importance of inputs? See ftp://ftp.sas.com/pub/neural/importance.html (this is a temporary URL subject to change).

Setiono, R. (1997a). Extracting rules from neural networks by pruning and hidden-unit splitting. *Neural Comput* **9,** 205-25.

Setiono, R. (1997b). A penalty-function approach for pruning feedforward neural networks. *Neural Comput* **9,** 185-204.

Specht, D. F. (1990). Probabilistic neural networks. *Neural Networks* **3,** 110-8.

Stahlberger, A. & Riedmiller, M. (1997). Fast network pruning and feature extraction by using the Unit-OBS algorithm. *Advances in Neural Information Processing Systems* **9,** 655-61.

Wu, C. H., Zhao, S., Chen, H. L., Lo, C. J. & McLarty, J. (1996). Motif identification neural design for rapid and sensitive protein family search. *Comput Appl Biosci* **12,** 109-18.

Glossary

ab initio Completely from the beginning, as opposed to pre-existing. An ab initio prediction is one that is constructed from theory, rather than from an empirical correlation.

accuracy A measure of how closely the predicted output of a neural network or other prediction/classification method matches the desired (target) value(s):

$$= \frac{\text{number of true positive predictions} + \text{number of true negative predictions}}{\text{total number of predictions}}$$

activation function A description of the action taken by a neural network unit or neuron in response to an input value or set of input values (see logistic function). Nonlinear activation functions are responsible in great part for the power and flexibility of neural networks.

adaptive resonance theory A set of neural network methods based upon neurophysiological principles. The basic adaptive resonance theory (ART) network is a two-layer feed-forward and feed-backward network. The first layer is the input/comparison layer and the second is the output/recognition layer. Input signals are exchanged back and forth between the layers in a cyclic fashion, hence the term resonance.

adaptive training A method of presenting input vectors to a neural network during training such that unlearned patterns are presented in inverse proportion to their occurrence in the training set. In ordinary back-propagation training, all patterns are presented equally.

alphabet A sequence alphabet is a set of characters, from which sequences are composed, like the DNA alphabet (A, T, G, C), or the set of 20 single-letter amino acid codes.

α-helix One of the major secondary structures in polypeptides. See secondary structure.

algorithm   A step-by-step problem-solving procedure, typically an iterative computational procedure for solving a problem in a finite number of steps·

amino acid   One of the approximately 20 building blocks that make up a protein molecule. Amino acids have a free amino group on one end and a carboxyl group on the other, and a range of possible side groups.

Amino acid residue   An amino acid in a peptide or protein linkage, as opposed to a free amino acid.

architecture   The structure of a neural network, including the number of layers, number of units within a layer, the connections between them and their activation functions.

ART network   See adaptive resonance theory.

artificial neural network   A (usually) digital algorithm that simulates some of the massively parallel functions of a biological neural network. This is accomplished by connecting small computational units (neurons) in such a way that the networks can be trained (or allowed to train themselves) to perform classification, pattern recognition and other artificial intelligence tasks.

axons   The part of a nerve that conducts signals away from the body of a nerve cell (neuron). Dendrites conduct signals toward the body of the nerve cell. In artificial neural networks these functions are simulated by connections of various strength (weights) between neuron units.

back-propagation   A popular method of training multilayer perceptrons by adjusting the strength of connections (weights) as a function of their effect on the output error. Error at the output layer is propagated back into the hidden units as part of this process. Multilayer perceptrons are occasionally called back-propagation networks.

base pair   Two nitrogenous bases (adenine and thymine or guanine and cytosine) held together by weak bonds. Two strands of DNA are held together in the shape of a double helix by the bonds between base pairs.

basis functions   A family of symmetric mathematical functions whose magnitude decrease with distance from their "center". These functions are usually multidimensional. The Gaussian function is the most commonly used basis function.

Bayesian techniques A method of training and evaluating neural networks that is based on a stochastic (probabilistic) approach. The basic idea is that weights have a distribution before training (a "prior" distribution) and another (posterior) distribution after training. Bayesian techniques have been applied successfully to multilayer perceptron networks.

benchmark data A standard collection of data sets that can be used for comparison of different techniques and architectures.

bias In statistical usage, bias is the error encountered in estimating a parameter, i.e. the difference between the true and estimated values. In neural networks, bias is an input unit with fixed output value that is connected to other neurons in the network. In practice, the value is usually 1.0 or −1.0. The weight between a bias unit (sometimes called a bias term) and a neuron is equivalent to the "constant" term in ordinary regression.

black box A hypothetical device with known input and output values and unknown and unobservable internal processing functions.

BLAST Basic Local Alignment Search Tool. One of the more widely used tools in genome informatics, for similarity searches of protein or DNA sequences.

bootstrapping A computationally intense method of training based upon repeated sampling, with replacement, from a sample of data. Particularly useful when training data is not abundant. The major use of bootstrapping with neural networks is to estimate generalization error. See cross-validation for a different method.

CAAT BOX An element of promoters lying upstream of transcription start sites. Binds with CCAAT-binding transcription factor.

cap site The site on a DNA template where transcription begins.

cascade-correlation One of the "network growing" algorithms that start with small networks and add complexity as needed to fit the training data.

cascading networks A set of interconnected neural nets, with the output of some networks being used as input to other networks.

causal importance One of the evaluation criteria to determine the importance of a given input unit (or variable) in a neural network: a measure of change in output as related to a change in input.

chromosomes Self-replicating genetic structures in cells which contain the DNA sequences that constitute genes. Prokaryotes have only one chromosome, eukaryotes have multiple chromosomes.

classification The process of deciding by some means (a neural network, for example) which of several groups or classes an object (say a protein or DNA sequence) is most similar to. In general, an application of vector mapping, that maps a vector (input pattern) into one of a discrete set of output states (classes).

clustering The process of grouping similar objects into a simpler object. For example radial basis function networks map similar training vectors into clusters which can be represented by a single function or central vector whose center and shape describe the similar vectors in a more parsimonious manner.

codon A group of three nucleotides that specify (code for) one of the amino acids during translation. See mRNA.

collinearity A statistical term describing variables (e.g. inputs to neural networks) that are highly correlated. Protein sequences can also be called collinear in that they may have similar linear orders and relative distances of amino acids. Training is complicated by colinearity.

competitive learning An unsupervised training method in which input vectors are mapped to a "competitive" layer. The neuron with the highest input in the competitive layer is the "winner". Weights to this neuron are then adjusted to increase the signal strength. This process is repeated over all input patterns iteratively until some stopping criterion is reached.

comparative genomics The comparison among genomes to gain insight into the universality of biological mechanisms and into the details of gene structure and function.

computational biology The field of study that aims to apply computer methodology to problems in biology. This includes the development of algorithms and other techniques that facilitate the understanding of biological processes.

confidence limits The endpoints of a confidence interval. A confidence interval will contain the true value of an estimated statistic (say the value of a weight, or the output of a neural network during training) a certain percentage of the time (typically 95%). For example, if repeated sampling from a training data set is used to train a neural network, 95% of the output values for a given pattern will be within the confidence interval. Bootstrapping and other sampling techniques are often used to estimate confidence limits for neural networks.

conformation The shape of a molecule - usually the tertiary structure of a protein.

consensus sequence A DNA sequence that is highly conserved in different species; or, the smallest common sequence that appears in homologous polynucleotides or proteins.

conserved sequence A base sequence in a DNA molecule (or an amino acid sequence in a protein) that has remained essentially unchanged throughout evolution.

contiguous map The alignment of sequence data from adjacent regions of the genome to effect a continuous sequence across a chromosome region; used as a strategy to help sequence the human genome. Contig map.

continuous Without gaps or discrete changes. Continuous variables may take on an infinite number of values within a given interval.

convergence A property of algorithms that will lead to a desired result (e.g. solution of a problem) in a finite number of steps. For example, the perceptron training algorithm is guaranteed to converge to a solution in a finite number of steps for a certain class of problems.

correlation coefficient A statistic with value between -1 and $+1$ that indicates the magnitude of correlation or relationship between two variables. A value of -1 indicates perfect negative correlation (an increase in one variable is associated with a decrease in the other); $+1$ indicates perfect positive correlation (an increase in one is associated with an increase in the other) and 0 no correlation (a change in one is not imply a change in the other).

cost function See error function.

counter-propagation A neural network with three layers of neurons: input, competitive and Grossberg (output) that is trained by a combination of unsupervised and supervised learning. The competitive layer is trained by a competitive learning algorithm, similar to Kohonen Self-organizing Maps. The final layer is trained in a supervised mode with a delta rule similar to what is used in back-propagation

cross-validation A method of training based upon dividing data into subsets of equal size, selecting all but one subset for training and using the remaining subset for testing. The process is repeated until all subsets have been excluded from training. Cross-validation is typically used to estimate generalization error. This is different from bootstrapping, where selection of subsamples is done randomly. An extreme version of cross-validation is the "leave-one-out" procedure where the subset held out for testing consists of only one case.

C-terminal In a polypeptide sequence, the amino acid residue which is connected to the end of the sequence by its amino group, leaving it with a free carboxy group.

curve-fitting A method of estimating parameters of a mathematical model or neural network to closely match a set of data. The model is "fit" to the data. Fit is determined by some error function. A synonym for regression.

data encoding The process of manipulating or extracting information from a set of data to make it suitable for input to a neural network. For a very simple example: a long protein sequence may be encoded by counting the number of occurrences of each amino acid; this results in a 20-element vector, one element for the count of each amino acid. In practice, this is perhaps the most complicated and crucial step in applying neural networks to genome informatics.

decision tree classifiers Programs or algorithms that perform classification by constructing decision trees or diagrams. By a series of decisions (each decision results in taking a specific "branch" to the next decision), objects are classified into two or more groups. Binary trees have dichotomous (yes/no) outcomes at each branch.

delta rule A rule for adjusting weights during training by an amount proportional to the difference between the predicted and target output values. The difference is called delta, abbreviated by the symbol Δ. Also called the Widrow-Hoff delta rule.

dendrites The part of a nerve that conducts signals toward the body of a nerve cell (neuron). Axons conduct signals away from the body of the nerve cell. In artificial neural networks these functions are simulated by connections of various strength (weights) between neuron units.

dense encoding Output encoding that involves only one output unit, i.e. has a binary (yes/no) result.

density functions Mathematical functions, which describe the distribution of random variables. The height of the function for a given value of a dependent variable is the probability density at that value. The area under the curve (the integral over all possible values) is 1.0. The area under the curve between two points is a probability. A common example is the Gaussian function.

derivative The slope of a curve (rate of change) at a point is the instantaneous derivative at that point. Technically, a function is said to be differentiable at x if and only if:

$$\lim_{h \to 0} \frac{f(x+h)-f(x)}{h} \quad \text{exists.}$$

If the limit exists, it is called the derivative of f at x. Functions that have a derivative over a range of values are called differentiable over that range.

diagonal matrix A matrix with values on the diagonal and all off-diagonal elements with the value 0.

differentiable See derivative.

dimensionality The number of elements in a vector or matrix.

direct encoding A method of encoding that converts each element (amino acid or nucleotide) into a vector. For example an amino acid in a sequence could be encoded as an indicator vector of 20 elements, all 0 except for one which has value 1 in the position reserved for that amino acid. Direct encoding preserves positional information, but cannot be encoded into a fixed length window (the size of the window varies with the length of the sequence).

direct search A table lookup method of searching based on pair-wise sequence comparisons.

discriminant components The variables or combination of variables used in discriminant analysis to help classify observations into two or more groups of similar objects (see classification).

distance measure A measure of the geometric distance between two points or vectors. See Euclidean distance. Also, in sequence analysis, a measure of similarity between two sequences. Two closely related sequences will have a small distance measure. See Hamming distance.

DNA   Deoxyribonucleic acid.   The molecular basis of the genetic code – a macromolecule formed of repeating deoxyribonucleotides (formed from 4 nitrogenous bases – adenine (A), thymine (T), cytosine (C) and guanine (G)). These bases form complimentary base-pairs: A-T and G-C on separate strands of the helical DNA molecule.

DNA sequence   The relative order of base pairs, whether in a fragment of DNA, a gene, a chromosome, or an entire genome.

domain   A discrete portion of a protein with a unique function. The overall function of a protein is determined by the combination of its domains.

downstream   Toward the 3' end of a DNA or other polynucleotide chain.

dynamic momentum rate   A momentum rate that is changed during training, rather than having a fixed value throughout.   Various algorithms for changing the momentum rate have been devised.

dynamic node architecture learning   A method of neural network training that allows the architecture (the number of layers, units within layers and connections between them) to be determined automatically.

dynamic programming   A recursive algorithm for minimization that works backward from a desired solution to the optimal answer according to some cost function.   In informatics, dynamic programming is used to perform optimal sequence alignments.

dynamic self-adaptation   A genetic algorithm to adjust the learning rate to the landscape generated by the error function.   The procedure has been used to improve the convergence rate compared to that of back-propagation.

encoding   The process of preparing information for presentation to a neural network. Important features (variables, functions of variables) may be extracted from "raw" data in the process of encoding. Also, in molecular biology, the DNA in exons is said to "encode" for specific proteins.

enhancer   A 50-150bp sequence of DNA that increases (i.e., enhances) the rate of transcription of coding sequences.

entropy A measure of randomness of lack of information in data.

epoch The complete set of training patterns (input vectors) for a given problem. Training proceeds in a stepwise fashion through each of the training patterns. When all patterns have been presented to the network, this constitutes an epoch. Learning algorithms, such as back-propagation, may update weight vectors after each pattern is presented or, more commonly, after each epoch.

error function A measure of performance for a neural network during training. Typically (see below), it is the squared difference between actual output values and target values (desired results), summed over all training patterns. The goal of training is usually to minimize the error function.

$$E = \sum_{i=1}^{n} (o_i - t_i)^2$$

where o_i is the output from the network for pattern i and t_i is the desired target value and n the number of training patterns. With more than one output neuron, another summation (over all output neurons) is added to the equation.

eukaryotes Organisms with well-developed subcellular compartments, including a discrete nucleus. Virtually all organisms except viruses, bacteria and algae are eukaryotes. See prokaryotes.

Euclidean distance A measure of the distance between two vectors. If X and Y are vectors of length n (with elements x_i and y_i) then the Euclidean distance between them is given by:

$$D = \|X - Y\| = \sqrt{\sum_i (x_i - y_i)^2}$$

evolutionary algorithm A training or search algorithm based on the concepts of evolution, i.e., mutation, recombination, reproduction and selection and fitness. Function values, strings, etc. are manipulated in some way and then tested for "fitness" by some cost function. Successful elements are selected, others eliminated. This is done in an iterative fashion until some prearranged criterion is met.

exon The part of a DNA sequence in a gene that contains codons that may be used for constructing an amino acid sequence. The region is "expressed" as opposed to intron regions which are not expressed. The regions are also said to "encode" an amino acid sequence.

expert system An artificial intelligence technique based upon rules generated by human experts. A computer program uses a series of "if...then..." tests to perform its tasks.

family classification Classification of biomolecules into large classes or families of molecules with similar structure and/or function.

family search/reverse search A database search method that is based upon comparisons of motifs, domains or family membership rather than pair-wise sequence comparisons.

FASTA A method of sequence similarity search to find sequences in a protein data library that are similar to a protein query sequence.

features Units of biologically relevant information in the sequence that may provide insight into their structure and function. Features may include individual components, counts of alphabetical letters and other mathematical functions of the sequences.

feature extraction The process of manipulating input sequences to create or discover features that may be useful for input into a classification or other neural network application.

feature map A matrix or other mathematical representation of summary features from a sequence.

feature representation The exact forms in which features extracted from sequences are encoded.

feed-forward In a neural network or other data-flow diagram, feed-forward indicates that information is passed in a sequential direction from layer to layer, starting with the input layer. This is opposed to feedback networks where information is also passed in a backward direction.

filter A device or algorithm that performs some mathematical operation(s) on input data before passing it on as output. See transfer function.

fit As a verb, fit means action taken in regression or neural network training to "learn" from a set of training data. As a noun, it is a measure of how closely the trained network output data matches the desired (target) data.

Fourier transform A mathematical method of breaking a signal (function or sequence) into component parts (for example, any curve can be approximated by the summation of a finite number of sinusoidal curves). In genome informatics, the Fourier transform of a sequence is used as a means of extracting information about the sequence into a more tractable, smaller number of features.

fully-connected All neurons (units) of one layer are connected to all neurons of the subsequent layer. Multilayer perceptrons, for example, are usually (but not necessarily) fully connected: all input units are connected to each hidden unit and all hidden units are connected to each output unit.

function In mathematics, a mapping (a rule of assigning for each value of one variable a unique value to another variable) of one or more variables into another. For example, $y = x^2$ is a simple mapping of a variable, x into a variable y which is the square of x; the function of x is x^2 in this case.

function approximation A method of estimating the approximate value of a function at certain points. For example, a sufficiently large neural network can be trained to approximate mathematical function, such as $\log(x)$, or $\sin(x)$. In these examples, the approximation is never exact, but can be made accurate to within a given amount of error.

function minimization A method used to reduce a function, such as an error function to an acceptably low value. Neural network training is nothing more than finding a set of weights that can minimize the value of the error function over a set of training patterns (input vectors).

functional analysis Analysis of the functions of proteins, rather than their structures.

fuzzy logic A form of logic that does not have discrete outcomes such as 0 (false) or 1 (true) but a distribution of values. Outcomes of operations may be expressed as probabilities.

Gaussian function A highly useful function named after mathematician Carl Friedrich Gauss. The familiar bell-shaped function is symmetric and has the property that its integral is 1. In statistics, a Gaussian distribution is called a "normal" distribution and has the familiar parameters mean (μ) and standard deviation (σ):

$$f(x) = \frac{1}{\sigma\sqrt{2\pi}} e^{-(x-\mu)^2/2\sigma^2}$$

GC-box A guanine and cytosine-rich sequence present in the promoter region of many genes; the binding site for a specific protein.

gene The fundamental physical and functional unit of heredity. A gene is an ordered sequence of nucleotides located in a particular position on a particular chromosome that encodes a specific product (i.e., a protein or RNA molecule).

gene expression The process of converting information within a gene into amino acid sequences and proteins or RNA sequences. The process includes transcription to mRNA and then translation into proteins.

generalization The neural network property of being useful (e.g. making accurate predictions) for a set of data that it was not trained with.

genetic algorithm See evolutionary algorithm.

genetic code The sequence of codons in mRNA that determines the sequence of amino acids in protein synthesis. The DNA sequence determines the mRNA sequence, which in turns dictates the amino acid sequence.

genetic map A linear map of the relative positions of genes along a chromosome. A linkage distance is determined by the frequency at which two gene loci become separated during chromosomal recombination. A linkage map.

genome The complete set of all genetic material present in the chromosomes of an organism; its size is generally given as its total number of base pairs.

genome informatics The systematic development and application of computing systems and computational solution techniques for analyzing data. Includes the development of methods to search databases quickly, to analyze sequence information and to predict structure and function from sequence data.

global minimum The minimum error function that exists in the parameter space. In neural network training, it is desired to find that exact set of weights that minimizes the error function. Unfortunately, for multilayer perceptrons, there is no guarantee that a global minimum exists or that it is unique. This is opposed to local minimum, where in a restricted region of the parameter (e.g. weight) space it is possible to find a minimum point.

glycosylation Post-transcriptional modification of a protein by the addition of a carbohydrate moiety.

GRAIL A gene prediction program.

Hamming distance The Hamming distance between two sequences of equal length is the number of character positions in which they differ. The Hamming Distance is a distance measure.

Heaviside function A mathematical function whose value is either 0 or 1, depending upon the magnitude of the input (independent variable). One of several so-called "thresholding" functions used in neural networks to transform weighted sums of inputs into a neuron into a binary output response.

heuristic Involves solving a problem or learning something by experimental, i.e. trial and error, methods. For example, there are no well-established methods for determining the number of hidden units in a neural network. Typically, units are either added or subtracted until the error function is affected dramatically.

hidden layers The inner layers of a neural network, not the input or output layers. If a black box analogy for a neural network is used, the layers inside the box, not visible from the outside, are the hidden layers.

hidden Markov model A probabilistic tool based upon states and transition probabilities between states; used multiple sequence alignment for example.

hidden units The neurons or units that make up hidden layers of an artificial neural network.

homology Similarity in DNA or protein sequences between individuals of the same species or among different species.

Human Genome Project (HGP) The collective name for several projects begun in 1986 to map and sequence the entire human genome.

hybrid system An artificial intelligence system that is the combination of several different types of systems, for example a neural network whose output is connected to a fuzzy logic system.

hydropathy index A measure of polarity of an amino acid residue; the free energy of transfer of the residue from a medium of low dielectric constant to water.

hydrophilic The property of an amino acid residue that has a charged or polar side chain that allows it to form hydrogen bonds. Hydrophilicity confers water solubility.

hydrophobic The property of an amino acid residue that makes it not mix freely with water. This property is quantified by a measure, its hydrophobic moment.

hyperbolic tangent function A sigmoidal function that maps inputs into a smooth differentiable function whose output values lie between -1 and $+1$. Used as an activation function for hidden and output units in neural networks.

importance Relative value of a neural network component to generalization. See causal importance and predictive importance.

independence The property of having no correlation or relationship; an important statistical assumption in many models.

indicator vector A vector of binary components with usually only one element having a value of 1 indicating either position or composition. For example, the DNA alphabet could be encoded into indicator vectors (1 0 0 0), (0 1 0 0), (0 0 1 0) and (0 0 0 1), with each vector indicating one of the four letters A,T,G or C.

indirect encoding A method of sequence encoding that is not a one-to-one mapping of a sequence, but an extraction of certain features from the sequence. As opposed to direct encoding, indirect encoding does not preserve positional information but has the advantage of being useful for unequally sized sequences.

informatics The study and application of computer and statistical metehods to the management of information. See genome informatics.

information theory   The mathematical field of study of how information can be encoded and measured. See entropy.

initial weights   The starting value of weights in a neural network, before training. Usually set to a small random number.

initiation codon   The codon, e.g., AUG, that signals the start of translation of a protein.

input encoding   See encoding.

input layer   The first layer in a network that usually plays the role of presenting input data to the hidden layers, i.e. the transfers functions have a constant value of 1.0.

input variable   One of the input elements or features in an input vector that is to be presented to a neural network: analogous to independent variables in a regression problem.

input vector   A vector or pattern of values to be presented to a neural network. In a classification setting, the input vectors contain the features thought to be helpful in performing the classification.

intron   A region of DNA in a gene that is not allowed to encode a protein sequence. See exon.

inverse folding   The protein folding problem is this: given an amino acid sequence what 3-dimensional structure will result? The inverse folding problem is: given a structure what sequences are compatible with this structure? This problem is also called the fold recognition problem.

jury   A committee or group of networks whose combined performance or decision is better than any single network's performance.

kernel regressions   A regression technique similar to radial basis function networks. Parameters for a set of kernel functions (see basis functions) are determine that best approximate the input data. A related method is the nearest-neighbor algorithm.

k-fold cross-validation   See cross-validation. Cross validation with k divisions of the input data.

K-means algorithm An iterative technique for • automatic clustering. The first step in a Kohonen self0organizing map algorithm.

knowledge extraction The process of learning from a trained network what the important discriminating data features are.

knowledge-based A system based on a database of information augmented by inference rules.

Kohonen self-organizing map An unsupervised learning method of clustering, based on the k-means algorithm, similar to the first stage of radial basis function networks. Self-organized maps are used for classification and clustering.

k-tuples Groups of elements in a sequence of size k. Counts of k-tuples, such as consecutive amino acid pairs (e.g. AA, AC, AD AE,…) are often used as features in a training input pattern.

Latent Semantic Indexing A method of information retrieval which takes advantage of implicit high-order structure in the association of terms • to improve the detection of relevant information.

learning. The process of minimizing the error function (by manipulating the weights in a network) for a given set of training data. The process by which a network learns to perform its approximation or classification tasks.

learning algorithm. The method by which a network is trained or learning occurs. There are two fundamental classes of learning algorithms, supervised and unsupervised.

learning rate A modification to the step size (change in weight) in a back-propagation training algorithm. A high learning rate forces large changes in weights, a low learning rate forces small changes in weights at each iteration. Some algorithms vary the learning rate dynamically. In the training rule: $w_{ij} \Leftarrow w_{ij} + \eta \delta_j o_j$, η is the learning rate.

learning vector quantization The name given to a Kohonen algorithm to iteratively construct modified training sets. The modified sets are then used in place of the original training sets.

leave one out An extreme form of cross-validation in which each training pattern is removed from a set of training patterns, the network trained and tested on the pattern left out. See cross-validation.

likelihood function. A probabilistic model used in training. The parameters of the model (e.g. weights of a neural network) are chosen to maximize the probability (the likelihood) of the output in response to the training data.

linear combination A linear combination of variables (elements, vectors or functions) x_1, x_2, x_3, x_4, ...is of the form $ax_1 + bx_2 + cx_3 + dx_3 +$....where a, b, c, d, etc. are real numbers.

linear discriminant analysis A statistical method of classification (discrimination) based on a linear discriminant, a linear function of the input vectors.

linear regression A technique of fitting a linear model to a set of data. As with other training situations, the object is to find values of the parameters that minimize an error function.

linearly separable Classification problems that can be solved with hyperplane decisions surfaces are linearly separable. For example, two groups of data that can be separated in a plane by a straight line are linearly separable. Perceptrons can only solve problems that are linearly separable, a major limitation.

local minimum A minimum of an error function in a small region of the parameter space; as opposed to a global minimum.

logistic function A mathematical function that is a smooth, differentiable approximation to a step function. It maps all input values to a number between 0 and one. The logistic function, one of the sigmoidal functions, is commonly used as the activation function for hidden and output neurons in multilayer perceptrons.

$$f(x) = \frac{1}{1 + e^{-\beta x}}$$

look-up table A method of function mapping by looking up matching values in a table to find the result (output) of the function.

LVQ See Learning Vector Quantization.

major histocompatibility complex The products of genes grouped together on a chromosome that determine whether transplanted tissues will be accepted or rejected.

mapping See function (in mathematics). In molecular biology, mapping is the determination of the physical location of a gene or genetic marker on a chromosome.

Markov process A stochastic process in which the next state of a system depends solely on the previous state.

matrix A table of values in rows and columns. Strictly, a rule that assigns to pairs of integers (rows and columns) a unique value for all such pairs in the matrix. A matrix of just one column or row is called a column or row vector.

memorization An undesirable results of training in which a network functions solely as a look-up table and performs no generalization. Each pattern in the training set is said to have been memorized.

minimal architecture The minimum number of layers and units in a network necessary to generalize satisfactorily.

mixture-of-experts model

MLP See Multilayer Perceptron.

moiety A fragment of a molecule, especially one that is an identifiable unit.

momentum term A term added to the delta rule for back-propagation training to restrict the magnitude of weight changes. In the training rule:

$$w_{ij} \Leftarrow w_{ij} + \eta \delta_j o_j + \alpha \Delta$$

α is a constant ($0<\alpha<1$) called the momentum term and Δ is the magnitude of the weight adjustment ($\eta \delta_j o_j$) made at the previous iteration.

motif A recognizable subsequence or substructure of a macromolecule (DNA, RNA or protein) that usually has functional significance.

mRNA Messenger RNA. In translation, the RNA that contains the coded information, as sequences of codons, for protein synthesis; mRNA serves as a template for protein systhesis.

multilayer perceptron A feed-forward neural network with an input layer of neurons, an output layer and one or more hidden layers between them. The activation functions for a multilayer perceptron are typically nonlinear. The back-propagation training method is commonly applied to multilayer perceptrons. See MLP.

multiple alignment A rectangular arrangement, where each row consists of one sequence padded by gaps, such that the columns highlight similarity/conservation between positions. A score is computed based on similarities and gaps.

multiple linear regression Linear regression with more than one predictor (independent) variable.

nearest-neighbor classification A method of classification an input point or vector by putting it into the class of its nearest neighbor, with "nearest" being defined by some distance measure.

negative predictive value The proportion of input patterns predicted to be negative by a network that are actually negative (true negatives). See sensitivity/specificity.

neighborhood size In Kohonen self-organizing maps, a grid of points is used for clustering. Points close to an important grid point, become a part of the cluster for that point. The radius of what is meant by "close" is the neighborhood size. The size of a neighborhood is one of the parameters of the network that must be determined.

network architecture See architecture.

network growing The processes of increasing the size/complexity of a neural network to enhance its ability to generalize. The opposite of network pruning.

network parameters Components of a network (e.g. connection weights) that must be modified during the training process to facilitate learning (minimize the error function)

network pruning The process of removing parts of networks with the objective of improving generalizability. For example, heuristically modifying the number of hidden units is a form of pruning.

network topology See architecture.

neural network See artificial neural network.

neuron The basic unit of a neural network layer. Typically artificial neurons perform a weighted sum of inputs and pass this to an activation function to compute the neuron output.

n-gram encoding See k-tuples.

node Typically used to denote a neuron or unit of a neural network. Points on a grid are also called nodes. See neuron.

nonlinear classification Classification using nonlinear functions of the input data.

Nonlinear regression A form of regression where the model is not linear.

normalize A method of data standardization to restrict the range of data values. A typical standardization technique is to subtract the overall mean value from each element in an input vector and divide by the standard deviation. Other techniques just scale the data to have values between 0 and 1.

N-terminal In a polypeptide sequence, the amino acid residue which is connected to the end of the sequence by its carboxy group, leaving it with a free amino group.

nucleic acid A large molecule such as DNA or RNA composed of nucleotide subunits.

nucleotide A subunit of DNA or RNA consisting of one of the bases (adenine, guanine, thymine or cytosine for DNA, adenine, guanine, uracil or cytosine for RNA), a phospate molecule and a sugar molecule (deoxyribose in DNA and ribose in RNA). Used as building blocks for genes and the genetic code. See DNA.

optimal brain damage A method of network pruning to reduce the number of layers, units or connections to facilitate better generalization. In spite of the novel name of the technique, it is based on a long-used technique in statistics based on the second derivative of the error function with respect to a given weight.

OR function A logical function that is true if and only if either of its two inputs is true.

orthogonal Perpendicular, or as expressed in vector form, two vectors are orthogonal if their inner products are 0. In statistical terms, orthogonal vectors are independent.

orthogonal least squares methods A type of linear modeling using orthogonal functions.

orthonormal A set of orthogonal vectors normalized such that their inner products have the value 1.

output layer The final layer of a feed-forward network that produces the output response of a network to input patterns.

overfitting See a related term, overtraining. Overfitting is often used to describe what happens with too many hidden layers and/or units in a network, for a given problem (i.e. the model is over-parameterized). The result is that generalization is affected.

overtraining The process of training a network to the point that generalization is not possible.

PAM Point Accepted Mutation. The PAM matrix is a frequency table representing substitution rates for closely related proteins at the particular *evolutionary distance* represented by multiple sequence alignments.

partially connected A neural network which does not have all the units (neurons) of one layer connected to its subsequent layer. Multilayer perceptrons are usually fully-connected.

pattern recognition The process of classifying an input pattern into one of a finite number of known classes or patterns.

PCA See principal component analysis.

peptidase An enzyme that cleaves the peptide bonds of proteins and peptides.

peptide A sequence of amino acids linked with peptide bonds. Arbitrarily, peptides have fewer than 50 or 100 amino acids, proteins have more.

peptide bond the amide bond formed between the amino and carboxyl groups of two adjacent amino acids in a peptide or protein.

perceptron A neural network with only two layers, an input layer and an output layer. The output layer has simple activation functions, typically a step function (Heaviside function). Perceptrons are guaranteed to converge to a solution for linearly separable problems, if the problems have a solution.

perceptron convergence theorem The theorem which proves that the perceptron learning rule will converge to a solution of a linearly separable, if a solution exists.

perceptron learning rule An iterative procedure for adjusting the weights of a perceptron neural network. See delta rule.

phylogenetic (evolutionary) analysis A study of the evolutionary relationships which are usually inferred from the similarities of amino acid sequences of contemporary proteins or of the base sequences of contemporary nucleic acids. A phylogenetic tree shows the divergence of contemporary species from their common ancestors and the branch points at which different species separated.

polarity The degree of electronic charge (polarization) in a molecule, in particular an amino acid residue. One of the features used in classification.

positive predictive value The proportion of input patterns predicted to be positive by a network that are actually positive (true positives). See sensitivity/specificity.

post-processing Conversion of the output from a neural network into a different, often more usable form.

prediction set See test data.

predictive importance One of the evaluation criteria to determine the importance of a given input unit (or variable) in a neural network: related to the change in generalization error when a unit is omitted or added.

preprocessing Manipulation of data before presentation to a neural network. See encoding.

primary structure The sequence of amino acid residues in a peptide or protein or the nucleotide sequence in a polynucleotide. See secondary structure.

principal component analysis Principal Components Analysis. A method used to reduce the dimensionality of a training set. The idea is to find functions (principal components) of the original data that can represent the same information in a more parsimonious form.

probabilistic neural network A special form of a radial basis function network such that each input pattern or vector has its own basis function. Typically the mathematical problem is to find the appropriate width parameter (e.g. standard deviation) for the radial basis functions. It is called probabilistic because the basis functions are usually probability density functions such as the Gaussian function.

probability There are many definitions of probability, the simplest, called the frequentist definition, is the ratio of the number of times a given outcome occurs to the number of possible outcomes. Probability depends upon the observed results, given a hypotheses (model) as opposed to likelihood which looks at the hypothesis, given the results. Probabilities have a value between 0 and 1.

probability distribution The mathematical distribution of a random variable. See density function.

processing units An elementary element, neuron, in an artificial neural network.

prokaryotes Organisms or cells lacking a discrete nucleus. See eukaryotes.

promoter A site or region on DNA that contains binding sites for RNA polymerase and proteins that regulate transcription.

propensity The likelihood of finding an amino acid residue in a certain secondary structure, e.g. glycine in an α-helix.

protein A large molecule composed of amino acid residues in a specific sequence by peptide bonds. Proteins are essential for the structure, function and regulation of cells, tissues and organs. See peptide. The linear arrangement of the amino acids is the protein sequence.

protein data bank (PDB) A large international repository of macromolecular data.

protein folding problem The problem of determining the tertiary structure of a protein, given its primary structure, i.e. its sequence of amino acid residues.

pseudoinverse A particular kind of matrix inverse that is identical to the ordinary inverse except when the matrix is singular; in such cases the pseudo inverse is the minimum least squares solution to the problem: Y = XB, where X is a matrix and Y and B are vectors.

QuickProp   A variant of back-propagation that uses a modification of the weight-decay rule. In some cases it has been shown to converge significantly faster than back-propagation.

radial basis function network   A neural network technique that consists of two steps during training and evaluation. Input vectors are mapped to a layer of basis functions in an unsupervised manner similar to the way Kohonen maps are trained. The output of the basis functions are then mapped to output nodes in a supervised manner similar to the method used for perceptrons (or with ordinary linear regression).

random sequence   A sequence whose elements appear to be in random, unpredictable order. Randomness is not well-defined, but its presence or absence can be evaluated by entropy measures and various statistical techniques.

random coil   One of the major secondary protein structures. Others include α-helices, β-sheets and β-turns.

reading frame   The register in which the translation apparatus senses the information coded within an mRNA molecule. As the code is a triplet, there are three possible reading frames.

real-valued   The property of being assigned the value of a real number, as opposed to a letter, or other symbol.

regression   **The procedure of fitting a mathematical model to data. Similar to training of a neural network.**

regulatory region or sequence   A DNA base sequence that controls gene expression.

residue   See amino acid residue.

restriction map   A physical map showing the locations of restriction enzyme recognition sites.

reverse turn A secondary structure of proteins in which the backbone turns sharply on itself. Also known as a β-turn or a hairpin bend.

ribosome-binding site The Shine-Dalgarno sequence of four to seven nucleotides that occurs in the leader section of an mRNA molecule and functions to properly orient the mRNA with the ribosome for protein synthesis.

RNA Ribonucleic acid. A macromolecule made of nucleotide sequences similar to DNA, but with uracil instead of thymine as a base and ribose as the backbone sugar instead of deoxyribose. RNA plays a major role in transcription and translation of the genetic code. There are several classes of RNA including mRNA, tRNA and rRNA.

rRNA Ribosomal RNA. A class of RNA found in the ribosomes of cells.

ROC curve Receiver Operating Characteristics curve. A graphical technique of evaluating the performance of a classifier: the plot of sensitivity by 1 – specificity (see sensitivity/specificity).

root-mean-square error See error function.

secondary structure The regular folding (conformation) of a protein in repeated patterns, e.g. α-helix, β-pleated sheet, β-turns and random coils. See primary and tertiary structure.

self organizing map A neural network that uses an unsupervised learning technique to automatically perform clustering and other tasks. See Kohonen self-organizing map.

sensitivity/specificity Calculations used to quantify the success or failure of a classification or prediction technique. Sensitivity is the proportion of truly positive outcomes that are predicted to be positive. Specificity is the proportion of truly negative outcomes that are predicted to be negative. This is illustrated by the following table and formulas.

Predicted Outcome	True Outcome (target value)		Total
	Positive	Negative	
Positive	a b	a + b	
Negative	c d	c + d	
Total	a + c	b + d	

Sensitivity = a/(a + c)   Positive Predictive Value = a/(a + b)

Specificity = d/(b+d)   Negative Predictive Value = d/(c + d)

sequence encoding   See encoding or data encoding.

sequence alignment   A process of comparing DNA or protein sequences. Typically this is done by aligning like residues and assigning a score based on matches and penalties for gaps.   Attention can be focused on local alignments or global alignments.

sequencing   The process of determining the linear sequence of nucleotides in a DNA/RNA molecule or amino acid residues in peptides/proteins.

sheet   A major class of protein secondary structures, one example is a β–sheet. See secondary structure.

side chain   The moiety of an amino acid residue in a protein, or of a free amino acid, that is attached to the α-carbon and is unique to each amino acid.

sigmoidal function   An "S"-shaped function used in neural networks to map weighted sums into smaller range of values. See logistic function and hyperbolic tangent function.

signal peptide   The N-terminal portion of a secretory or membrane protein that assists it across the membrane of the rough endoplasmic reticulum, where it is synthesized, but is cleaved from the protein even before the synthesis of the protein is complete.

simulated annealing   A form of function minimization (learning) based upon analogy to thermodynamics.   Network weights are adjusted randomly, with decreasingly smaller changes as training progresses, i.e., as the network "anneals".

singular value decomposition   A method of reducing the dimensionality of training data based upon finding a vector ranking features in order of importance and choosing some number of the most important features. Related to principal component analysis.

softmax function   A method used to make the sum of the multiple outputs from a neural network equal to one, so that the outputs are interpretable as posterior probabilities. In practice it is a multiple logistic function of the form:

$$o_i = \frac{e^{net_i}}{\sum\limits_{j=1}^{k} e^{net_j}}$$

For k output nodes, where o_i is the output from unit i and net_i is the input (weighted sum of inputs) into unit i.

sparse encoding The use of two output units to represent a binary result rather than one unit (dense encoding).

specificity See sensitivity/specificity.

splice acceptor junction A segment of DNA at the 3' end of an intron and the 5'-end of the next exon that facilitates excision and slicing reactions.

splice donor junction A segment of DNA at the 3'-end of an exon and the 5'-end of an intron that facilitates excision and slicing reactions.

splicing A processing procedure in which two exons, or coding sequences, are joined together in a eukaryotic mRNA after excision of the intervening introns; used to produce a mature RNA molecule.

spline function A function used to approximate a set of data piecewise over a specified interval. The pieces, splines, match each other closely (usually in value and derivatives) at their overlapping points. Spline functions simulate the effect of a draftsman's spline, a tool for drawing complex curves.

squashing function An activation function that maps all values into a small range, usually between 0 and 1. Examples of squashing functions are the Heaviside function, logistic function and other sigmoid functions. The input values are "squashed" into the restricted output range.

start/stop codons See initiation or termination codons.

step function A function whose value is 1 for all input values above a threshold value and 0 below. See Heaviside function.

stiff differential equations   Differential equations with widely varying rate constants. Like neural networks, their solution depends upon careful selection of step sizes.

stopping criteria.   A rule used to terminate the iterative training process for neural network learning or function minimization. To prevent overtraining, the stopping criteria may not be based solely upon the error function; for example performance on a validation set is often used to stop training.

supersecondary structure   The arrangement of elements of secondary structures in a protein.

supervised learning   A method of network training that uses a set of correctly classified examples. The value of an error function is minimized.

SVD   See singular value decomposition.

synapse   The junction between two interacting neurons.

synaptic weight   A weight term in a neural network that simulates the strength of connection between two neurons.

target value   In supervised learning, a correctly classified set of patterns is used. The correctly classified values are called target values. The network output for each pattern is compared to its target value in an error function.

TATA box   An adenine and thymine rich consensus sequence of eukaryotic DNA that is part of the promoter region and is upstream of a gene. The binding site for RNA polymerase.

term weighting   A method used to vary the relative importance of a feature, say in an input pattern, by multiplying the value of the feature by some weight term.

termination codon   A codon (polynucleotide triplet) that signals the limit of protein synthesis.

termination condition   A synonym for stopping rule in an iterative training process.

tertiary structure   The unique three-dimensional structure of a macromolecule. See secondary structure and primary structure.

test data A set of data for evaluation of the results of training. The network has not "seen" or presented with the test data during its training. See validation data and training data.

threshold functions A step function or other function that has a different output above and below a threshold value. See Heaviside function.

tiling algorithm A method used to optimize the network topology by adding hidden units and layers dynamically.

tolerance A name given to a pre-defined size of the error function that is used to terminate training iterations. A measure of how much error will be "tolerated".

training Network learning or more rigorously: minimization of an error function by manipulation of the weights in a network.

training data Date used for training a neural network. A set of input patterns to be presented to the network during training. For supervised training, the input patterns are accompanied by the correct classifications for each training pattern. See validation data and test data.

transcription The first step in gene expression: the copying of a template from DNA to mRNA. See translation.

transfer function The mathematical relationship between the input to a unit and its output. See function or mapping.

translation The process of converting the template in mRNA into a polypeptide. tRNA molecules carry the appropriate amino acids to ribosomes for assembly into peptides.

transmembrane domain A feature of most intrinsic proteins of plasma or vesicular membranes; a polypeptide sequence of about seven residues if β-sheet, up to 22 residues if α-helix, that connects extracellular to intracellular domains, joined by extended polypeptides on the cytoplasmic and external or vesicular sides.

tRNA Transfer RNA. See translation.

unit A component of a neural network layer. Synonym: neuron or artificial neuron.

unsupervised learning A method of training neural networks that doesn't depend upon target values (correctly classified examples). The training algorithm automatically performs the necessary functions. See Kohonen self-organizing map.

upstream Toward the 5' end of a DNA or other polynucleotide chain.

validation data Data used during training of a neural network to independently evaluate the ability of the network to generalize. Some algorithms stop training when the performance of the network with validation data has stopped improving significantly, independent of the value of the error function. Training data is used for the function minimization; validation data is used to test the efficacy of the network during training; and, test data is data to be applied to the network after training has been completed.

validation error The error encountered when applying validation data to a network during training.

vector A collection of related numbers. In pattern recognition, an input vector is the set of features (a pattern) to be presented to a neural network for classification. A row or column of a matrix.

weight In neural networks, a weight is a number representing the strength of a connection between two neurons. The parameters to be manipulated during error function minimization.

weight decay A method of reducing the weights of a neural network during training by adding a penalty term to the error function. Apparently smaller weights are important to generalization.

weight matrix The set of all the weights in a neural network. This includes the weights from the input layer to the first hidden layer, weights between hidden layers and weights from the last hidden layer to the output layer.

weight sharing A method of reducing weight terms in a neural network by forcing several neuron to neuron connections to have the same weight.

weighted average The average of several numbers obtained by multiplying each of them by a factor (a weight) and dividing the weighted sum by the sum of the weights.

For example, if x_1, x_2 and x_3 are the numbers to be averaged and their respective weights are w_1, w_2 and w_3:

$$weighted\ average = \frac{w_1 x_1 + w_2 x_2 + w_3 x_3}{w_1 + w_2 + w_3}$$

weighted sum   The sum of several numbers multiplied by a weight term.   The denominator in the weighted average formula.   In neural networks, the input to hidden and output neurons is the weighted sum of all the outputs from the previous layer multiplied by their respective weight terms: for unit i this is sometimes called net$_i$ or net input to unit i.

window size   In examining sequences, sometimes it is effective to look a small subsets of the sequence as if looking through a window or template.   Some techniques of sequence analysis use a sliding window, in which the window is advanced by one position after each calculation.

zinc finger   A structural motif which appears in a DNA-binding protein with amino acid residues that serve as ligands for zinc.   A metal-binding motif.

Author Index

All authors are cited and indexed in reference sections.
All authors are indexed in the text but only the first two authors are cited.

Subject Index